SUPPORTING SELF-DIRECTED LEARNING IN SCIENCE AND TECHNOLOGY BEYOND THE SCHOOL YEARS

While much has been written about science education from pre-K through to postgraduate study, interaction with science and technology does not stop when schooling ends. Moving beyond scholarship on conventional education, this book extends the research and provides an original in-depth look at adult and lifelong learning in science and technology. By identifying the knowledge and skills that individuals need to engage in self-directed learning, the book highlights how educators can best support adult learners beyond the years of formal schooling. Through case studies and empirical analysis, the authors offer a research-based exploration of adults' self-directed learning and provide tools to support adults' learning experiences in a wide range of environments while being inclusive of all educational backgrounds.

Léonie J. Rennie is Emeritus Professor of Science and Technology Education in the School of Education at Curtin University, Australia.

Susan M. Stocklmayer is Emeritus Professor of Science Communication and Founder of the Australian National Centre for the Public Awareness of Science at The Australian National University, Australia.

John K. Gilbert (1949–2020) At the time of writing this book, he was Professor Emeritus of Science Education at The University of Reading, and Visiting Professor of King's College London, UK.

TEACHING AND LEARNING IN SCIENCE SERIES

Series Editor: Norman G. Lederman

Responsive Teaching in Science and Mathematics
Edited by Amy D. Robertson, Rachel Scherr, David Hammer

Reconceptualizing STEM Education
The Central Role of Practices
Edited by Richard A. Duschl, Amber S. Bismack

Supporting K-12 English Language Learners in Science
Putting Research into Teaching Practice
Edited by Cory Buxton, Martha Allexsaht-Snider

Reframing Science Teaching and Learning
Students and Educators Co-developing Science Practices In and Out of School
Edited by David Stroupe

Representations of Nature of Science in School Science Textbooks
A Global Perspective
Edited by Christine V. McDonald, Fouad Abd-El-Khalick

Teaching Biology in Schools
Global Research, Issues, and Trends
Edited by Kostas Kampourakis, Michael Reiss

Theory and Methods for Sociocultural Research in Science and Engineering Education
Edited by Gregory J. Kelly, Judith Green

Supporting Self-Directed Learning in Science and Technology Beyond the School Years
Léonie J. Rennie, Susan M. Stocklmayer, John K. Gilbert

For more information about this series, please visit:
www.routledge.com/Teaching-and-Learning-in-Science-Series/book-series/LEATLSS

SUPPORTING SELF-DIRECTED LEARNING IN SCIENCE AND TECHNOLOGY BEYOND THE SCHOOL YEARS

Léonie J. Rennie, Susan M. Stocklmayer, and John K. Gilbert

Routledge
Taylor & Francis Group
NEW YORK AND LONDON

First published 2019
by Routledge
52 Vanderbilt Avenue, New York, NY 10017

and by Routledge
2 Park Square, Milton Park, Abingdon, Oxon OX14 4RN

Routledge is an imprint of the Taylor & Francis Group, an informa business

Library of Congress Cataloging-in-Publication Data
A catalog record for this title has been requested

ISBN: 978-1-138-35325-1 (hbk)
ISBN: 978-1-138-35326-8 (pbk)
ISBN: 978-0-429-43430-3 (ebk)

DOI: 10.4324/9780429434303

Typeset in Bembo
by Taylor & Francis Books

CONTENTS

ILLUSTRATIONS

Figure

Tables

PREFACE

This book is about how adults learn the science and technology they need to know in order to deal with issues in their everyday lives, and how other people can help them to learn. Science and technology are fundamental to the way we live and how we interact with our environment. They are also fundamental to the way things are done in a cultural sense and they interact continually with our social lives. But not all people are comfortable with science and technology and frequently, as issues arise in their lives, they need to further their knowledge and understanding to cope more easily with those issues. In this book we explore a range of stories about how adults coped with different circumstances and how they accessed various resources and other people to learn the science and technology they needed to help them to cope. But let's begin with a story about the ways that science and technology permeate everyday life.

Not long ago, Clarissa, an antique dealer, picked up her loupe to examine an elegant brooch comprising several clear, white stones set in a whitish-coloured metal. The owner watched, hopefully. "Yes", said Clarissa, "these are diamonds, and old ones, too. You can tell by the way they are cut." She pointed out the culet, the small facet on the base of the diamond, to the excited owner: "See, there is a bottom facet on each diamond, so from the top, the stone looks like it has a tiny hole in it. These days, diamonds are cut so that the stone is less deep and there is no culet. That way, although the stone is smaller, it reflects more light and there is more sparkle". "Are the diamonds set in silver?" asked the owner. "No, the metal is platinum," replied Clarissa, "That's why it looks so white. Silver would tarnish and the inside parts of the setting would be grey." "So it's not worth as much?" said the owner, disappointed. Clarissa smiled. "Actually, platinum makes a better setting because it is more durable, but it has a higher melting point than silver so it is harder to work. Platinum was used more frequently in jewellery from around the 1900s, so that helps us to date the setting. This brooch is over 100 years old, and it is very valuable."

The social habit of adorning oneself with pretty things is as old as the human race, and styles and fashions varied according to the materials available and the customs and technical skills of the time. Diamond faceting began to exhibit elements of symmetry in the 15th century, but with cutters aiming to retain the largest possible stone because diamonds are beautiful, rare, and therefore expensive. Diamond is the hardest natural gemstone so faceting is difficult, but it has a very high refractive index and high dispersion, so it looks its best when cut to display maximum internal reflection and dispersion. By the turn of the 20th century, gem-cutting lathes and polishing techniques had improved to facilitate great precision in faceting. During the early 20th century, better understanding of the science of light reflection and dispersion enabled calculation of how many facets should be cut, and at what angles, to get the best effect. The modern round brilliant cut diamond has 58 facets, but the 58th facet at the base of the cut gem, the culet, is usually omitted so that light at right angles to the table facet cannot pass straight through, thus avoiding the "hole" effect in old-cut stones. Pre-20th century stones were cut deeper (to preserve weight) and, as diamond is brittle, including the culet removed the "sharp point" so the stone was less likely to break if knocked on its base. Silver was commonly used as a setting in jewellery because it is a soft metal and easy to shape; gold was preferred because it didn't tarnish, but it was rarer and more expensive. Platinum is harder and also rare, but more difficult to work due to its high melting point, so it was not until the techniques and tools for working with it were developed that its use in jewellery became more common.

The story of jewellery is, like many other histories, the story of the interplay between science, technology, engineering, and mathematics. The skills and techniques of cutting and faceting of gem materials were learned initially by trial and error, but once the science of light behaviour was understood and the underlying mathematical relationships between refraction, reflection, and dispersion were calculated, leaps forward were possible. Similarly, the design and manufacture of jewellery-making equipment, and the designing and making of the jewellery itself, developed first from trial and error, then with greater sophistication as the chemical and physical properties of the materials became known, and tools were developed to facilitate precision in construction. And, of course, the relevant science, technology, and mathematics interact with aesthetic, social, and cultural customs. Humans find gemstones to be beautiful and being durable, they are used for adornment. Because taste and preferences vary within cultures and over time, jewellery can be dated by the cut of the gems and the design of the setting. For example, the brooch Clarissa was assessing may well have been of the art nouveau style if it dated from around 1900. The quality of the gem, the elegance of its setting, and its rarity both in terms of uniqueness and craftsmanship, are the determinants of its value.

Clarissa's story as an antique dealer is also a story of interplay, but here the interplay is between the various sources of learning that developed the skills that enabled her to identify the nature and quality of the stones and their setting, to date it, and subsequently to assign it a monetary value. Clarissa was now an experienced dealer who had come a long way in her literacy about science and technology since leaving school. As a school student, she had no firm career goal in mind and took a general science course. After leaving school, she began employment as a sales assistant in a jewellery store, soon becoming fascinated by the jewellery itself and wanting to know much more about it. She learned about the trade by working as an assistant to a jeweller who also did valuations, then found her specific interest lay in antique pieces. She sought experience by looking at and researching jewellery design through reference books, and talking to other dealers. Early on, she realised she needed a much more detailed understanding of gems and their characteristics so she completed a formal course in gemmology to ensure accuracy in the identification and assessment of the quality of the stones. Later, she took a short course in jewellery design, enabling her to appreciate the skills needed to create and manufacture a setting, and also a course on evaluation. Now, she keeps in touch with others in her trade, and uses the internet to stay abreast of current prices at auctions. However, she willingly admits that she is not up-to-date with the science of new synthetic gemstones or treatments to improve the colour and clarity of natural stones. She doesn't meet these new variants often in the antique trade, but knows she will need to take an updating course on this topic to counter the increasing number of replica pieces that are coming into the market. Her learning never stops!

Clarissa's learning journey is an illustration of the many different ways people come to learn about the science and technology they need to know about the things that matter to them in their everyday lives. Because many adults have left school without a firm career pathway, or because circumstances require them to change their employment later on, most of this learning happens outside of, and beyond, schooling. Falk and Dierking (2002) elaborated this theme by explaining how people continue to learn throughout the various stages of their lives. Although in some cases inklings of one's formal science schooling are reawakened, essentially many of the things we learn as adults result from learning by ourselves, seeking sources of information, pestering people who know more than we do, and gradually building up the working knowledge that enables us to do what needs to be done to make our lives better, for example, or to pursue a hobby interest, or just satisfy our curiosity about something old or new.

In this book we retell a number of stories of learning that people have told to us. These are the stories of adults who, for one reason or another, did not have the requisite background in science and technology to respond to personal issues that arose in their lives. The stories explore how these people searched for relevant and understandable information to address their need to know and how they were able to sort out what was useful to them and what was not. All of these people were motivated to respond to their need to know, and by identifying the commonalities in their patterns of exploration and describing the different ways these people went about learning, sometimes over a lifetime, we will distil the overarching features that determine how people learn in a self-directed way. Our aim is to provide a framework that informal science educators can use to assist people to learn; not to teach them formally, but to guide them in less formal ways to fulfil their need to know.

In general terms, this is a book about adult education, but in specific terms, this is a book about adult learning of science and technology. Why are we focusing on science and technology? Because science underpins what we know about how the world works, so having relevant knowledge about science enables us to understand it and, together with the associated technological artefacts, to shape it to help resolve related issues. As adults become aware of and affected by emerging new technologies, they begin to realise that the science they may have learned at school has become less relevant to the immediate needs of their everyday life. Keeping up with what is happening around them requires new learning about science and technology. Further, our aim in this book is not to explain how to teach, but to tease out ways to support adults who are motivated to learn because they have chosen to respond to an issue related to science and technology that has come into their lives. In some of our case stories this is a medical issue, in other cases the issues relate to employment, to hobbies, or the need to assist another person to learn about topics in science and technology. In all cases these adults were self-motivated to learn, and they sought help from a range of different resources. For the most part, these resources were not formal courses, but rather sources of information from what is often called the informal domain. Our focus is to determine how to help people to find and better understand the bits of science and technology that they need to know to resolve, or to come to terms with, their issue.

This is also a book about building literacy in science and technology. Fundamentally, being able to self-direct one's learning means increasing one's skills in scientific and technological literacy (surprisingly similar to what has come to be known as 21st century skills). Consequently, we begin this book with a discussion of what is meant by science, by technology, and by STEM (Science, Mathematics, Engineering, Technology), an acronym that has increasingly permeated discussion in this field. We will explore the meaning of these terms, why they are important, and suggest answers to questions about how much we need to know about them. We move then to Chapter 2 where we explore learning, what it means and what is especially pertinent to the learning of adults. The core of the book then follows: Four illuminative chapters of personal case stories, each in the words of their authors, and clustered according to a common issue or motivation. Then, to illustrate diversity in learning styles, Chapter 7 provides a case study that describes how a variety of adults tackled a common assessment task. From our case stories we synthesise how our authors accessed resources and the features that made them useful or not, and what characteristics our learners required to enable them to be effective, self-directed learners. Chapter 8 deals with general resources and Chapter 9 focuses specifically on new media. In Chapter 10 we draw together the threads from these two chapters and our exploration of adult learning in Chapter 2 in order to tease out ways to best support adult learners. Finally, in Chapter 11, we look back and review how our case story authors became more scientifically literate and how education, in general, might be structured more effectively to promote understanding about science and technology.

Reference

Falk, J. H., & Dierking, L. D. (2002). *Learning without limits: How free-choice learning is transforming education*. Walnut Creek, CA: Altamira Press.

ACKNOWLEDGEMENTS

This book began in John's wondering how adults learned about science when they had a real need to do so. Where did they start? What did they do? Were there ways that they could be supported in their learning? John asked Sue and Léonie for their ideas. We assumed that we could go to some well-thumbed books about adult learning and find out but, to our surprise, most of the books gave us a wealth of information about how to teach adults, but not very much about how adults learn, particularly when they were alone in their learning task, and especially if information were needed in a hurry. These books gave us an idea about a possible approach to adults learning by themselves, usually referred to as self-directed learning, but it was in the context of taught material.

We decided we should ask some adults how they went about learning when they had a personal need to know something about science and technology, so we did. All of the resulting stories were different, but the common threads we uncovered revealed that there was indeed a lot to know about how adults learned, particularly if we wanted to support their personal learning. This book is the result.

We thank Naomi Silverman, then of Taylor & Francis in New York, for her early encouragement and belief that we could write this book. We thank Norm Lederman and later, Karen Adler and Emmalee Ortega, for their support as we put the manuscript together. We are grateful to the anonymous reviewers who made valuable suggestions as the manuscript was taking shape.

Most of all, we thank our wonderful authors who, willingly and generously, took the time to write about their personal learning journeys. We greatly appreciate their assistance for, without them, there would be no book, and John would have no answers to his questions.

Our authors, listed in the order they appear in the book, are Ana Afonso, Penny Dufty, Mary Hooper, Richard Rennie, Michael Marsh, Tina Jarvis, Dr Paulette M McManus, Paul Hill, Elizabeth Warren, Tiki Swain, Kristen Mahoney, Warren Hurley, Hugh Thacker, Ketan Joshi, and Keith Beuman.

We owe them much. Thank you for giving us answers and helping us to suggest ways forward for other learners.

Addendum to the Acknowledgement

As we noted in the introduction to our acknowledgements above, this book was initiated by John Gilbert, and it is a testament to his ever-inquiring mind and persistent pursuit of ways to help people to learn. It was a four-year task, managed by three people in three locations, and it succeeded because John carried the three of us on a fascinating, and significant, academic journey.

Now, sadly, there are just two of us. Sue and Léonie remember many long, enjoyable, and often irreverent, conversations with John. We talked about science and science education, and formal and informal approaches to learning about science. We were always amazed at the nimbleness of John's mind, the breadth of his knowledge, and his incredible ability to synthesise massive amounts of information and come up with the pith that mattered.

This book received very supportive reviews and it has succeeded because of John's commitment to helping people to learn. Even though he was its major instigator, he did not want to be first author, preferring to remain as our leader in the background.

We have chosen to make this book open access as a tribute to John Gilbert, our inspirational colleague and mentor, and our much-missed friend.

1

WHAT ARE SCIENCE AND TECHNOLOGY?

It is not too much of a stretch to say that science and technology (and more recently STEM – science, technology, engineering, mathematics) have become buzzwords of our times. There are so many definitions of each that it is very easy to get confused, especially if the aim is to find a clear-cut, unambiguous definition. We can begin to get an understanding of what these terms mean by looking at some dictionary/thesaurus definitions and teasing out the commonalities. Table 1.1 presents the results of a search for "science" gleaned from a range of dictionaries, texts, and web-based sources. It shows four common meanings for science and also gives examples of what might count as science for each meaning. Table 1.2 provides the results of a similar search for the meaning of "technology".

Overall, two things stand out from Tables 1.1 and 1.2: Science is systematic; its purpose is to build dependable knowledge and understanding of the physical and natural world based on observation and experiment. Technology involves the development and manufacture of artefacts and the application of tools and systems to meet human needs.

It is easy to see why technology is sometimes considered to be the application of the knowledge of science, and sometimes it is, but it is more appropriate to describe science and technology as interrelated, continuously interacting together. A meteorologist might observe the weather, but the observations are far more useful for predicting weather if tools (such as thermometers, hygrometers, and satellite imagery) are used to extend and quantify those observations.

Perhaps surprisingly, a search of various learned scientific societies' websites found little that defined what was meant by science. One exception was the United Kingdom Science Council, who also "found that definitions of science were not readily available, and were not easily accessible on the Internet" (UK

DOI: 10.4324/9780429434303-1

TABLE 1.1 Common meanings of science with examples

Common meanings	*Examples of science for this meaning*
The systematic study of phenomena in the physical and natural world, especially by using observation and experiment	Identifying the factors that control the growth of anti-biotic resistant bacteria, or of tomatoes
A branch of science, the study or knowledge of a particular area of the physical or natural world	Geology, chemistry
A systematically organised body of knowledge about a particular subject	The anatomy of a horse, or the life of a star
An activity that is studied or performed methodically	Making meteorological observations or documenting plant species

TABLE 1.2 Common meanings of technology with examples

Common meanings	*Examples of technology for this meaning*
The development and application of artefacts, devices, machines, and techniques for manufacturing and productive processes to meet human needs and desires	Design and manufacture of a screwdriver, a smart phone, or a chocolate éclair
A methodology for the application of technical knowledge or tools	The process of making accurate meteorological recordings, or stitching a straight seam on a garment
Equipment, machines, and systems considered as a unit	A thermometer, a steam engine, or a classification key for plants
The sum of all practical knowledge in a particular area	Knowledge about how to manufacture and use textiles, or to breed terriers

Science Council, n.d.). In 2009 the Council published its own definition of science so that it was clear what was meant "when it [the Council] talked about sound science and science based policy" (http://www.sciencecouncil.org/definition).

> Science is the pursuit and application of knowledge and understanding of the natural and social world following a systematic methodology based on evidence.
>
> *(http://www.sciencecouncil.org/definition)*

The United Kingdom Science Council's definition of science is quite consistent with our earlier analysis and so gives some endorsement to it, but it has two additional features. First, the definition includes "the pursuit and application of knowledge and understanding", whereas application of knowledge has more often been viewed as the province of technology. Second, science is described as concerning both "the natural and the social world", an addition that takes the topics of science study to a much broader perspective, because it embraces human behaviour and cultural contexts. Further, because technology is a response to human needs it also is strongly shaped by human behaviour and cultural contexts.

It is a short step to realising why science and technology are so important in our everyday lives. Most readers will possess a smart phone, for example, a mobile digital communication device that is also a computer, TV, radio, GPS, camera, and a portable library for books, music, and video, as well as a console for games and, depending on the apps downloaded, a monitor of personal health, well-being, and many other things besides. Such a device is clearly a technological artefact, but no more so than a pencil, a fork, or a toothbrush. Of course, there is considerably more science research behind the mobile device than the fork, for example, although the latter has existed in various forms for millennia.

The Importance of Science and Technology

It is now commonplace for school curriculum documents to argue the importance of science and technology in people's everyday lives, and for the need to understand, or at least not be afraid of, science and technology. One example is the vision that underpins the United States-based K-12 Framework for Science Education (NRC, 2012), where it is stated that a compelling case can also be made that understanding science and engineering, now more than ever, is essential for every American citizen. Science, engineering, and the technologies they influence permeate every aspect of modern life. Indeed, some knowledge of science and engineering is required to engage with the major public policy issues of today as well as to make informed everyday decisions, such as selecting among alternative medical treatments or determining how to invest public funds for water supply options. In addition, understanding science and the extraordinary insights it has produced can be meaningful and relevant on a personal level, opening new worlds to explore and offering life-long opportunities for enriching people's lives. In these contexts, learning science is important for everyone, even those who eventually choose careers in fields other than science or engineering (NRC, 2012, p. 7).

Another example comes from a report on citizenship to the European Commission (2015). In discussing why science education matters, it was argued:

> Knowledge of and about science are integral to preparing our population to be actively engaged and responsible citizens, creative and innovative, able to work collaboratively and fully aware of and conversant with the complex challenges facing society. It helps us to explain and understand our world, to guide technological development and innovation and to forecast and plan for the future. It introduces citizens to an important part of our European culture.
>
> *(p. 14)*

The message from both of these documents is consistent with rhetoric from almost every country: Our students and our citizens need to know about science and technology because they are important in their lives and a part of their culture. This message is not new; for over a century there have been calls for the public to be more knowledgeable about science, along with the practical view that people will live more fulfilling lives if they understand their science and technology dominated world. The arguments for this usually draw on an economic imperative, based in the belief that a better educated public will create a more prosperous nation, and a democratic or civic perspective that the public should be involved in decision-making about significant socioscientific issues and be supportive of government funding in science. There is also a cultural argument: Science and technology are part of our culture and cultural heritage, and should therefore be part of our world view.

These are strong and important arguments, but they have been contested. For example, the implication that all citizens should be able to have informed input to decision-making about policy in science-related matters has long been in dispute. The 1920s debate between Walter Lippmann (an influential journalist) and John Dewey (a leading philosopher and educator) about the capability of the public to participate constructively in public policy is a well-known example. Simplistically, their arguments revolved around human nature and democracy and, of particular interest here, whether the ordinary citizen can attain sufficient knowledge to participate in an informed way in decision-making about public policy, and whether in a democratic society the public should have a role in those decisions. Again simplistically, Lippmann's view was that the public voted for their decision-makers (their government) who were then responsible for making policy decisions based on guidance from experts. Thus the public had no direct input, and Lippmann believed this to be reasonable because the "truth" was obscured from citizens by the way the media reported it, and the human mind was not capable of understanding the subtleties required. Dewey's position was more democratic. He believed the public's view should be heard by government and also by the experts, suggesting a two-way communication between the public and experts, somewhat akin to recent exhortations for scientists to engage with the public.

Much has been written on this debate; for example, DeCesare (2012) discussed the arguments in terms of fundamentally different interpretations of knowledge, whereas analysis by Whipple (2005) focused on communication and democratic participation. Feinstein (2015) advanced a perspective about the Lippmann–Dewey debate that was more aligned with science education, to see what might be learned "about science, education, and democracy" (p. 146). He provided a simplified summary of Lippmann's and Dewey's views (see Feinstein, 2015, figure 1 on p. 158), but of course the matter is hardly simple. Wisely, Feinstein did not conclude with a directive to science education, but he teased out three significant issues. He referred to the need to pay attention to how people interpret information about science, how that information is reshaped by various media, and what platforms are available for the "public" to engage with scientists. All three of these issues will be visited (and often revisited) in the chapters that follow, but first we return to the importance of science and technology and what the everyday person might need to know about these disciplines.

Scientific Literacy and the Public Understanding of Science

In the 1950s, the terms *science literacy* and *scientific literacy* began to be used to describe what people understand about science but there was little clarity about what these terms actually mean, especially for the everyday citizen compared with people who are working in science. It is worth following some of the arguments in order to come to a position that will enable us to move forward in our thinking about how adults might be helped to learn about science and technology.

More than four decades ago, Shen (1975) proposed three kinds of scientific literacy needed by citizens. He referred first to *Consumer* scientific literacy, the level of knowledge that enables adults to shop for essentials like food, medicines, and other consumer goods. Second, *Cultural* scientific literacy describes citizens' understanding of science as a way of understanding the world, compared with other ways of knowing. Third, *Civic* scientific literacy denotes the knowledge needed to understand and deal with issues and arguments relating to scientific public policy.

Two decades later, when Shamos (1995) wrote about *The Myth of Scientific Literacy*, he criticised contemporary definitions of "true" scientific literacy as describing an unattainable goal in which all citizens could be educated to have a knowledge and understanding of science sufficient to be able to make independent judgements on socioscientific issues. He suggested that something more attainable would be

> (a) having an awareness of how the science/technology enterprise works, (b) having the public feel comfortable with knowing what science is about, even though it might not know much about science, (c) having the public understand what can be expected from science, and (d) knowing how public opinion can best be heard in respect to the enterprise.
>
> *(p. 229)*

Both Shen (1975) and Shamos (1995) were focusing on people going about the business of their daily activities, but also having some awareness of the science and technology in the world around them. This seems to be a more realistic approach to scientific literacy and the way the public might engage with science and technology, rather than assuming that all people can have sufficient knowledge and understanding to participate constructively in making decisions about science-related policy.

The arguments about the importance of science and technology led to a perceived need to measure how much the public actually knows about science, and over the last decades there have been regular attempts to survey the populations of various countries to obtain a measure of public understanding of science. Examples of this include the National Science and Engineering Indicators in the United States (National Science Board, 2018), and the European Commission's Eurobarometer surveys of public opinion (http://ec.europa.eu/public_opinion/index_en.htm). These surveys are based on responses to true–false questions of science-related factual knowledge. A core body of these questions derive from earlier work by Durant, Evans, and Thomas (1989) and Miller (1983), and include statements such as "Electrons are smaller than atoms". The 2018 report of the National Science Board Science and Engineering Indicators (National Science Board, 2018, table 7–1) found that correct responses to this particular item ranged from 27% to 58% across Canada, China, European Union, India, Israel, Japan, Malaysia, Russia, South Korea, Switzerland, and the United States (not respectively). However, it is difficult to see why it is necessary for every person to know this in order to pursue their everyday interests. Given the widespread over-prescription of antibiotics, it is probably more important for every person to know that the correct response to another statement, "Antibiotics kill viruses as well as bacteria", is "false". The National Science Board (2018) reported correct responses on this question of between 16% and 56% over the same list of countries.

Perhaps not surprisingly, given their content, most of these measures suggest wide-spread, low levels of public science understanding that are consistent over time. As a consequence, the general public has sometimes been described as illiterate about science and can hardly be blamed for thinking that science is something beyond their ken. However, this conclusion seems unfair because it is linked to results from instruments of questionable relevance. Stocklmayer and Bryant (2012) discussed the failings of these instruments and from their research found that not only are the instruments deeply flawed and open to question, but scientists themselves have often failed to answer the questions "correctly". Sjöberg (2015) has described some of the history and problems in defining public science literacy, its conflation with the term public understanding of science, and confusions and ambiguities relating to both, particularly concerning ways that they might be measured. Another problem is that such measures also seem to focus on what the public might know, and therefore what they do not know, a view of public understanding termed the "deficit model" (see Wynne, 1993). Efforts have also been made to try to find some kind of comparative global measure of the public understanding of science (see, for example, Bauer, Shukla, & Allum, 2012). Further confusion arises because the public is not one homogeneous body, it is heterogeneous and diverse, with many different "publics" depending on how one defines them (Bauer, Allum & Miller, 2007).

An overall consequence of these ambiguities is disparagement of the notion of scientific literacy itself, particularly in terms of the implied expectation that all citizens should be sufficiently informed to be involved in decision-making about socioscientific issues, some of which, like climate change and renewable energy, are both complex and controversial. Information about such issues is frequently portrayed negatively in the news media by focusing on disputes deriving from people having different interpretations of similar but incomplete scientific evidence. Further confusing the issue are the competing agendas of government and big business, with various social, cultural, economic, and political values supporting conflicting interpretations of how best to take action for the public benefit.

Finding a Meaning for Scientific Literacy

Finding agreement about what scientific literacy means is difficult, as Roberts (2007) pointed out in his masterly analysis: "there is a veritable deluge of definitions for SL" (p. 729). Roberts entitled his chapter "Scientific Literacy/Science Literacy" and throughout the chapter he used the abbreviation "SL" to mean both together. Indeed, some authors use the two terms interchangeably, while others might use one or the other term consistently but appear to write about the same thing. We draw attention to the crux of Roberts' approach by referring back to the "two legitimate but potentially conflicting curriculum sources: science subject matter itself and situations in which science can legitimately be seen to play a role in other human affairs" (p. 729).

As a heuristic device, Roberts (2007) proposed "two *visions* of SL that recently have come to represent the extremes on a continuum" (p. 730, original emphasis). Roberts named them Vision I and Vision II, indicating that a vision was broader than a definition, and explained them this way: "Vision I gives meaning to SL by looking inward to the canon of orthodox natural science, that is the products and processes of science itself" (p. 730); "Vision II derives its meaning from the character of situations with a scientific component, situations that students are likely to encounter as citizens" (p. 730). Roberts suggested that a curriculum drawing essentially from the products and processes of science exemplified Vision I, whereas a curriculum that was more outward-looking, by starting from situations or contexts in which science is important, exemplified Vision II. It is important to note that by making connections with science in the everyday, Vision II curricula still require disciplinary knowledge of science. The Visions are not mutually exclusive, as Roberts pointed out: "Vision II subsumes Vision I but the converse is not necessarily so" (p. 768).

We find this a helpful distinction in our quest to understand how adults learn about science in their everyday lives, well beyond their school science curriculum. If pressed to distinguish between science literacy and scientific literacy, we would agree with Roberts and Bybee (2014), who wrote that a Vision I perspective focusing inward on science more appropriately describes science literacy, whereas the broader Vision II perspective, with its emphasis on scientific situations that citizens face, describes scientific literacy. In our subsequent discussions we will therefore use the term scientific literacy, except where a distinction must be made because, for example, an author we cite refers to science literacy. So what meaning for scientific literacy can help us to move forward?

Feinstein (2011), writing in the context of science education, referred to

> the endless definition of and rationales for science literacy … it has come to mean everything and nothing. … What can be done to revitalise science literacy, to take it beyond the realm of politically useful slogans and make it into a goal that is both realistic and worthy?
>
> *(p. 170)*

Feinstein used the term science literacy, and in citing Roberts (2007) he referred to Vision I and Vision II as "answering different educational questions" (Feinstein, 2011, p. 170). Nevertheless, Feinstein implicitly championed a Vision II perspective because he claimed that "a truly useful version of science literacy must be connected to the real uses of science in daily life" (p. 168). In a vein that resonates with our view, Feinstein suggested that "we can salvage science literacy – make it into a meaningful educational goal instead of a mere slogan – by redefining it according to research on the actual uses of science in everyday life" (p. 183). Feinstein argued for a convergence between science education and public engagement with science; that science literate people "have learned to recognise the moments when science has some bearing on their needs and interests and to interact with the sources of scientific expertise in ways that help them achieve their own goals" (p. 180). This is in contrast to the earlier approaches to measure the public's scientific literacy that tended to focus more on knowing science, as exemplified by the factual questions used to survey public understanding of science.

Focusing on science that has bearing on people's everyday lives seems to be a sensible way to move forward and find a meaning for scientific literacy that can underpin the approach taken in this book.

What Does It Mean to be Literate in Science and Technology?

In general terms, we might think of the goal of any sort of literacy as "to provide people with the tools to participate intelligently and thoughtfully in the world around them" (Pearson & Young, 2002, p. 3). In terms of science and technology, then, what does this mean? How might we describe the scientifically literate person, or the technologically literate person? Are the skills required different? Are they the same? Do they overlap? For the most part, definitions for these things have been framed in the context of school education, but people spend only a small proportion of their lives in school. Therefore we need to ensure that the definitions fit adults, and are descriptive of the tools needed to participate intelligently and thoughtfully in science-related and technology-related matters in the world around them.

Here is a definition for scientific literacy developed in the context of schooling some time ago for a report to government on the status and quality of science education in Australian schools.

> Scientifically literate people are interested in and understand the world around them; engage in the discourses of and about science; are sceptical and questioning of claims made by others about scientific matters; are able to identify questions, investigate and draw evidence-based conclusions; and make informed decisions about the environment and their own health and well-being.
>
> *(Based on Rennie, Goodrum, & Hackling, 2001, p. 485)*

This definition was developed from an extensive review of international literature of the time, including *Project 2061: Science for All Americans* (AAAS, 1989) in the United States, which gave rise to the *National Science Education Standards* (National Science Council, 1996), the *Beyond 2000: Science Education for the Future* (Millar & Osborne, 1998) for the Nuffield Foundation in the United Kingdom, and the definition used in developing the OECD/PISA program (OECD/PISA, 1999). It is a good starting point in a practical sense because it describes the kinds of things that people might do that demonstrate that they are scientifically literate. Further, it has no inherent boundaries and so is applicable in a life-long sense.

Technological literacy can be defined in a way that has a similar structure by drawing from the literature in technology education, especially from Pearson and Young (2002) in the United States, Black and Harrison (1985) and Barlex and Pitt (2000) in the United Kingdom, Gardner, Penna, and Brass (1990) in Australia, and the curriculum documents from a number of other countries.

> Technologically literate people understand the designed world, its artefacts, systems and the infrastructure to maintain them; have practical skills in using artefacts and fixing simple technical problems; are able to identify practical problems, design and test solutions; recognise risks and weigh costs and benefits associated with new technologies; can evaluate, select and safely use products appropriate to their needs; and contribute to decision-making about the development and use of technology in environmental and social contexts.
>
> *(Rennie, 2003, p. 35)*

A focus on the differences between science and technology as disciplines can be reflected in the different purposes of their practitioners – generating and testing new knowledge and theoretical understanding in science, and producing tools, systems, and procedures to meet human needs – and also their products – explanatory theories in science and tools and systems in technology. There are clear parallels in the processes described by each definition. These parallels are demonstrated in Table 1.3 by employing a framework articulated by Pearson and Young (2002), who identified three dimensions of technological literacy: knowledge, capability, and ways of thinking and acting.

Fundamentally, when scientific and technological literacy are viewed in terms of the three dimensions of knowledge, capabilities, and ways of thinking and

TABLE 1.3 Parallels between scientific and technological literacy

Dimension	*Scientifically literate persons*	*Technologically literate persons*
Knowledge	Are interested in and understand the world around them	Use and understand the designed world, artefacts, systems, infrastructure
Capability	Engage in discourses of and about science	Have practical hands-on skills and fix simple technical problems
	Are able to identify questions, investigate and draw evidence-based conclusions	Are able to identify practical problems, design and test solutions, and evaluate results
Ways of thinking and acting	Are sceptical and questioning of claims made by others	Recognise risks, weigh costs and benefits
	Make informed decisions about the environment and their own health and well-being	Evaluate, select, and safely use products appropriate to their needs
		Contribute to decision-making about the development and use of technology

Source: Rennie, 2003, p. 39

acting, there is an underpinning similarity and that is why science and technology, and participation in both, are so frequently coupled together, at least in the minds of adults, if not in the school curriculum.

Where Does STEM Fit?

The increased usage of the acronym STEM (science, technology, engineering and mathematics) in school curricula (for example, Framework for K–12 Science Education [NRC, 2012] in the United States, and the National STEM School Education Strategy [Education Council, 2015] in Australia) and in discussion about shortages of workers in science and technology (for example, European Commission [EU Skills Panorama, 2015] in Europe) suggests that concerns about science, or about science and technology, have been overshadowed by a broadening of the focus to include engineering and mathematics. But what is meant by reference to STEM in these contexts? Based on an international review, Siekmann and Korbel (2016) concluded:

> [STEM] has grown to be an umbrella term for a variety of concepts, classifications and initiatives pertaining to not only learning and working in science and technology-related disciplines but to a nation's social contract and productivity. … The STEM acronym is often used in place of a more appropriate or precise term, thereby leading to misunderstandings and its disconnection from related initiatives, as well confusion in the media and public policy.
>
> *(p. 17)*

Clearly, greater precision is required. Most school curriculum documents tend to treat the components of STEM separately and meanings more precise than the source of the acronym "science, technology, engineering, and mathematics" are hard to find, although the term "STEM skills" is used repeatedly. What are STEM skills? Siekmann and Korbel (2016) believed that "STEM skills and knowledge are inter-disciplinary in nature", and that the integration of STEM's component disciplines is necessary to "enhance people's competency in work and/or life and more generally respond to social demands on technology" (p. 45). This, argued Siekmann and Korbel, requires a strong emphasis on critical and creative-thinking skills. Balka (2011), then president of the School Science and Mathematics Association (SSMA), pointed out that the "organisation's view has not been one of four silos (Science, Technology, Engineering, Mathematics), but one of true integration" (p. 7). He provided "a commonly used" definition of STEM literacy as "the ability to identify, apply, and integrate concepts from science, technology, engineering, and mathematics to understand complex problems and to innovate to solve them" (p. 7). Creative thinking is inherent in this view of STEM skills.

At the post-school level more complex definitions of STEM skills can be found. For example, the Skills Panorama Glossary (European Commission, 2018) defines STEM skills as "skills expected to be held by people with a tertiary-education level degree in the subjects of science, technology, engineering and maths". More specifically, they are defined as

> numeracy and the ability to generate, understand and analyse empirical data including critical analysis; an understanding of scientific and mathematical principles; the ability to apply a systematic and critical assessment of complex problems with an emphasis on solving them and applying the theoretical knowledge of the subject to practical problems; the ability to communicate scientific issues to stakeholders and others; ingenuity, logical reasoning and practical intelligence.
>
> *(EU Skills Panorama, 2015, p. 1)*

Expressed this way, STEM skills represent a very high level of literacy in science, technology, engineering, and mathematics. There is no country in the world where the majority of the adult population has "a tertiary-education level degree in the subjects of science, technology, engineering and maths". Indeed, the EU Skills Panorama previously noted that "A fifth of adults in Europe have only basic levels of literacy, while a quarter have only basic numeracy. Nearly half of adults have only basic levels of technological proficiency" (EU Skills Panorama, 2014, p. 1).

There is enormous variation of such skills within the general populace, and it would be more useful to have statements that might be viewed as aspirational, rather than unachievable. Fortunately, Bybee (2013) has described STEM literacy in a much more accessible way. In his view, a STEM literacy that applies to everyone, refers to an individual's

- knowledge, attitudes, and skills to identify questions and problems in life situations, explain the natural and designed world, and draw evidence-based conclusions about STEM-related issues;
- understanding of the characteristic features of STEM disciplines as forms of human knowledge, inquiry, and design;
- awareness of how STEM disciplines shape our material, intellectual, and cultural environments; and
- willingness to engage in STEM-related issues and with the ideas of science, technology, engineering, and mathematics as a constructive, concerned, and reflective citizen.

(p. x)

Clearly, this is a more realistic view for the everyday citizen. The definitions developed and teased out in Table 1.3 are very appropriate in this sense. Most people could say "yes, I can do that" to at least one of the statements, more literate people can say "yes" to more than one. Are similar statements needed for engineering and mathematics to account for the second half of the STEM acronym, that is, literacy in engineering and mathematics?

In terms of engineering, a recent development in the United States is instructive. The National Assessment Governing Board (NAGB) requested that the National Assessment of Educational Progress (NAEP) include a means of assessing literacy in technology and engineering (NAGB, 2014) as part of the current STEM curriculum initiative. This meant that technology and engineering had to be separated and defined, but it was found this was not easy to do. After considerable consultation, technology was defined as "any modification of the natural world done to fulfil human needs or desires" (p. 3), and engineering as "a systematic and often iterative approach to designing objects, processes, and systems to meet human needs and wants" (p. 3). These definitions consider technology essentially as artefacts or systems, and engineering as the process of designing those artefacts or systems. However, in terms of literacy, NAEP found it to be more sensible to combine these into a single definition of Technology and Engineering Literacy, and this was defined as "the capacity to use, understand, and evaluate technology as well as to understand technological principles and strategies needed to develop solutions and achieve goals" (p. 3). Consideration of this definition for technological and engineering literacy reveals that it is covered satisfactorily by the earlier definition for technological literacy, as demonstrated in Table 1.3.

Mathematical literacy also has its history of definition and redefinition, but in school curriculum terms, it is usually referred to as numeracy, and this is also how it is treated in the definition of STEM skills from the EU Skills Panorama (2015), discussed earlier. Adults who are not involved in highly mathematical activity need only a level of numeracy that allows them to know what numbers they need and what to do with those numbers; to calculate area, to decide how much of whatever they need, such as the amount of medicine to dose a child or the amount of fertiliser to put on a garden, the appropriate ratio of fat to flour in a recipe, and so on. For our purposes it seems unnecessary to consider mathematical literacy in any depth beyond this. For the adult learners referred to in this book, mathematical needs are likely to be covered by basic skills in numeracy.

For completeness, and before moving on, we mention STEAM (Science, Technology, Engineering, Arts, Mathematics), an acronym that has gained some currency in relation to school curricula over the last decade. If the term "arts" is interpreted to include the fine arts, manual arts, language arts, physical arts, and liberal arts, rather than just "art", then the breadth of the humanities is being linked with science, technology, engineering, and mathematics. Suddenly, the STEAM curriculum is unbounded! There is no argument that the major problems facing humanity today require tackling by the application of STEM skills, knowledge, and understanding in the context of cultural, social economic, and political values. In other words, creativity, effective communication, and ethical thinking underpin any solutions. We argue that a STEAM curriculum is simply one that teaches STEM in real world contexts (a point reiterated in Chapter 11). Indeed, Siekmann and Korbel (2016), who had a particular focus on the vocational education sector, took the view that, "Principally, STEM learning is a multi- or interdisciplinary approach to learning, in which academic concepts are coupled with real-world lessons to make connections between school, community, work, and business" (p. 17). Such a curriculum would benefit the later learning of adults and for them, as later chapters will reveal, acronyms such as STEM and STEAM are irrelevant: Their focus is on addressing issues in the context of their own world, a world that is entirely transdisciplinary. This book is about learning the science and technology they need to do that.

What Science and Technology Do People Need to Know?

This discussion of science, technology, scientific literacy and technological literacy has so far ignored the question of how much science and technology people need to know and understand. How scientifically and technologically literate do they need to be? Obviously, those with a career that is based upon science, technology, engineering, or mathematics need the specific set of knowledge, capabilities, and ways of thinking and acting that enables them to pursue that career. Those of us who do not have such a career can still benefit from a level of scientific and technological literacy, because it is helpful in our daily lives and enables us to understand considerably more of what is happening around us. We suggest that approaching the statements in Table 1.3 as aspirational, together with some number skills, is enough for the person without a career in science and technology. As noted earlier, the "public" comprising everyday citizens is heterogeneous and diverse, so having "enough" scientific and technology literacy will be variable and dependent on each person's life circumstances. It will be "enough" if it meets the nature of their needs and motivation to learn.

However, life circumstances do not stand still, but are ever changing. What of those citizens whose life circumstances change, for whatever reason, and suddenly, what they do know and understand about science and technology is not enough to cope with this change? These people are the focus of this book. They are the people who want to know or understand more about particular issues that are important in their lives. What can they do? What help can they access? What resources are available? How can best use be made of them? What assistance do these people need to find and make best use of those resources? How can we help them to learn to do this?

Where Can Adults Learn About Science and Technology and How Can We Help Them?

The answer to the first part of this question is very broad. One avenue is through courses in science and technology that are designed for adults, but as we will show in the next chapter, only a minority of adults pursue this formal route to learning. Instead, the majority of adults seek resources in the informal domain, sources like libraries, museums, friends, the Internet, and in their workplace. A decade ago, the Center for Advancement of Informal Science Education (CAISE) explored the "landscape" to map out the "network of informal science entities" that were available to the public "for free-choice STEM across their lifetimes and throughout the day" (Falk, Randol & Dierking, 2008, p. 1). Their map was impressive, ranging from informal education institutions like science centres, museums, and zoos, through media to various adult organisations and science writers (Falk et al., 2008, p. 12). CAISE updated this report in 2016, noting the new resources created over the previous ten years as the infrastructure in the informal science education field grew (Bell et al., 2016). In a word, there are myriad opportunities for learning science and technology in the community, and the case stories in Chapters 3 to 6 will illustrate how these resources are used.

In answering the second part of the question above, we emphasise that helping people learn does not necessarily mean teaching them. Adults who need to know are usually adults willing to learn by themselves; they are self-motivated, self-directed learners. To understand how to help them we need to think about how people learn, and also explore how people are motivated to learn and how they help themselves to learn. In the next chapter we provide basic theoretical frameworks about learning, particularly adult learning, and examine the psychological underpinnings of motivation and how people determine their own learning. In subsequent chapters, we present case stories to demonstrate the experience of people who, for various reasons, have sought to learn more science and technology to solve problems or to meet challenges that arose in their lives. These case stories will allow us to draw conclusions and guidelines about helping other adults to learn in a self-directed way.

References

American Association for the Advancement of Science (AAAS). (1989). *Science for all Americans*. New York: Oxford.

Balka, D. (2011). Standards of mathematical practice and STEM. *The Math–Science Connector*, Summer, 6–8. Stillwater, OK: School Science and Mathematics Association.

Barlex, D., & Pitt, J. (2000). *Interaction: The relationship between science and design technology in the secondary school curriculum*. London: Engineering Council.

Bauer, M. W., Allum, N., & Miller, S. (2007). What can we learn from 25 years of PUS survey research? *Public Understanding of Science*, 16(1), 79–95.

Bauer, M. W., Shukla, R., Allum, N. (Eds.). (2012). *The culture of science: How the public relates to science across the globe*. New York: Routledge.

Bell, J., Falk, J., Hughes, R., Hunt, G., Parrish, J., Ruffin, M., … Troxel, G. (2016). *Informal STEM education: Resources for outreach, engagement and broader impacts (A report by the Center for Advancement of Informal Science Education CAISE)*. Retrieved from http://www.informalscience.org/sites/default/files/CAISE_Broader_Impacts_Report_2016_0.pdf

Black, P., & Harrison, G. (1985). *In place of confusion: Technology and science in the school curriculum*. London: Engineering Council.

Bybee, R. W. (2013). *Case for STEM education: Challenges and opportunities*. National Science Teachers Association. Retrieved from http://ebookcentral.proquest.com/lib/curtin/detail.action?docID=1416112

DeCesare, T. (2012). The Lippmann-Dewey "debate" revisited: The problem of knowledge and the role of experts in modern democratic society. *Philosophical Studies in Education*, 43, 106–116.

Durant, J. R., Evans, G. A., & Thomas, G. P. (1989). The public understanding of science. *Nature*, 340, 11–14.

European Commission. (2015). *Science education for responsible citizenship* (Report to the European Commission of the expert group on science education). Brussels: Author.

European Commission. (2018). *Skills Panorama Glossary*. Retrieved from https://skillspanorama.cedefop.europa.eu/en/glossary/s

Education Council. (2015, December). *National STEM School Education Strategy: A comprehensive plan for science, technology, engineering and mathematics education in Australia*. Retrieved from www.educationcouncil.edu.au

EU Skills Panorama. (2014). *Adult Learning Analytical Highlight*, prepared by ICF GHK and Cedefop for the European Commission. Retrieved from https://skillspanorama.cedefop.europa.eu/sites/default/files/EUSP_AH_AdultLearning_2.pdf

EU Skills Panorama. (2015). *STEM Skills Analytical Highlight*, prepared by ICF and Cedefop for the European Commission. Retrieved from https://skillspanorama.cedefop.europa.eu/sites/default/files/EUSP_AH_STEM_0.pdf

Falk, J. H., Randol, S., & Dierking, L. D. (2008). *The informal science education landscape: A preliminary investigation*. Retrieved from http://www.informalscience.org/sites/default/files/2008_CAISE_Landscape_Study_Report.pdf

Feinstein, N. (2011). Salvaging science literacy. *Science Education*, 95(2), 168–185.

Feinstein, N. (2015). Education, communication, and science in the public sphere. *Journal of Research in Science Teaching*, 52(2), 145–163.

Gardner, P. L., Penna, C., & Brass, K. (1990). Technology and science: Meanings and educational implications. *Australian Science Teachers Journal*, 36(3), 22–28.

Millar, R., & Osborne, J. (1998). *Beyond 2000: Science education for the future* (The report of a seminar series funded by the Nuffield Foundation). London: King's College London, School of Education.

Miller, J. (1983). Scientific literacy: A conceptual and empirical review. *Daedalus*, 112(2), 29–48.

National Assessment Governing Board (NAGB). (2014). *Technology and Engineering Literacy Framework for the 2014 National Assessment of Educational Progress*. Retrieved from http://permanent.access.gpo.gov/gpo44685/ed005353p.pdf?ck=51

National Research Council (NRC). (2012). *A Framework for K-12 Science Education: Practices, crosscutting concepts, and core ideas*. Washington, DC: The National Academy Press.

National Science Board. (2018). *Science and engineering indicators: 2016–1*. Arlington, VA: National Science Foundation. Retrieved from https://www.nsf.gov/statistics/2018/nsb20181/data/tables

National Science Council. (1996). *National science education standards*. Washington, DC: National Academy Press.

OECD Programme for International Student Assessment (OECD/PISA). (1999). *Measuring student knowledge and skills: A new framework for assessment*. Paris: OECD.

Pearson, G., & Young, A. T. (2002). *Technically speaking: Why all Americans need to know more about technology*. Washington, DC: National Academy Press.

Rennie, L. J. (2003). Reassessing the role of literacy in technology education. In D. Fisher & T. Marsh (Eds.), *Proceedings of the Third Conference on Science, Mathematics and Technology Education* (Vol. 1, pp. 33–47). East London, South Africa: Key Centre for School Science and Mathematics, Curtin University of Technology, Perth, Western Australia.

Rennie, L. J., Goodrum, D., & Hackling, M. (2001). Science teaching and learning in Australian schools: The results of a national study. *Research in Science Education*, 31, 455–498.

Roberts, D. A. (2007). Scientific literacy/science literacy. In S. K. Abell & N. G. Lederman (Eds.), *Handbook of research on science education* (pp. 729–780). Mahwah, NJ: Lawrence Erlbaum.

Roberts, D. A., & Bybee, R. W. (2014). Scientific literacy, science literacy, and science education. In N. G. Lederman & S. K. Abell (Eds.), *Handbook of research on science education* (Vol. 2, pp. 545–558). New York: Routledge.

Shamos, M. H. (1995). *The myth of scientific literacy*. New Brunswick, NJ: Rutgers University Press.

Shen, B. J. (1975). Scientific literacy and the public understanding of science. In S. Day (Ed.), *Communication of scientific information* (pp. 44–52). Basel, Switzerland: Karger.

Siekmann, G., & Korbel, P. (2016). *Defining 'STEM' skills: Review and synthesis of the literature: support document 2*. Adelaide, South Australia: National Centre for Vocational Education Research. Retrieved from https://files.eric.ed.gov/fulltext/ED570655.pdf

Sjöberg, S. (2015). Public science literacy measures. In R. Gunstone (Ed.), *Encyclopedia of science education* (pp. 808–812). Dordrecht, The Netherlands: SpringerReference.

Stocklmayer, S. M., & Bryant, C. (2012). Science and the public: What should people know? *International Journal of Science Education, Part B*, 2(1), 81–101.

Whipple, M. (2005). The Dewey-Lippmann debate today: Communication distortions, reflective agency, and participatory democracy. *Sociological Theory*, 23(2), 156–178.

Wynne, B. (1993). Public uptake of science: A case for institutional reflexivity. *Public Understanding of Science*, 2, 321–337.

2

HOW DO ADULTS LEARN SCIENCE AND TECHNOLOGY?

In the previous chapter, we explained our understanding of science and technology and the need for adults to have access to ways to learn about science and technology beyond schooling. This chapter begins with an overview of adult education courses in science and technology, revealing evidence of limited provision. To identify avenues to increase participation in such learning, we then review the main approaches to adult learning to arrive at potentially fruitful ways of approaching the task by individuals. The success of such approaches will depend on acquiring suitable motivation and displaying self-determination in respect of the chosen task. Four broad contexts are identified as possible routes for the acquisition and display of such attributes. Lastly, the procedures adopted to acquire and analyse case stories of examples of such contexts are summarised. The case stories will form the basis of Chapters 3, 4, 5, and 6.

To What Extent Do Adults Learn Science and Technology?

In general, adults in most countries are not greatly engaged with science. They are engaged with the products of science and technology in a functional sense but not necessarily with an underlying understanding of how they are produced. There is ample evidence that, after the completion of schooling, little further "learning" of science occurs and much of what was learned in school is forgotten. In this book, we look at examples of adults who, for various reasons, have engaged with science and technology from "a need to know" basis. How did they set about this learning, given their lack of a useful scientific education?

It is well recorded that school science does not often result in lifelong learning (see for example, Fensham, 2008; Fensham & Harlen, 1999). Yet the educational policies of most governments suggest that featuring science and technology in the

DOI: 10.4324/9780429434303-2

school curriculum is vital for the educational pursuit of scientific literacy (see Chapter 1). It is not surprising that the immediate outcomes of the very substantial investments in teaching science and technology are monitored in varying detail at the national level in many countries. However, it is certainly surprising that the more fragmentary engagement with these subjects at the level of general adult education, that is, by the great majority of the population, is not considered as being of equal importance. For example, in the United Kingdom, the only evidence of official interest in the adult learning of science is subsumed in an annual survey of the whole field of learning.

In a recent annual survey by the National Institute of Adult Continuing Education of England and Wales (NIACE, 2015) a sample of 5,000 adults was asked if they had undertaken learning of any type in the last three years. Learning was defined to mean

> practising, studying or reading about something. It can also mean being taught, instructed or coached. ... Learning can also be called education or training. ... It can be full time, or part time, done at home, at work, or in another place like a college. Learning does not have to lead to a qualification. *(p. 6)*

Although the NIACE survey explicitly stated that it was intended to embrace "a wide range of formal, non-formal and informal learning", it is not clear how this range was either perceived or embraced by the respondents. Nevertheless, extracting the figures for the first cohort that is clearly post-school and post-university, the 25- to 34-year-olds, shows that 46% had undertaken some learning in the previous three years, with that proportion decreasing for older age groups. Most learners were drawn from high socio-economic classes; most were employed in the professions; the great majority (78%) would study for work or career reasons. Can we assume that, as science and technology have a high salience in employment today, a significant proportion of that 46% of 25- to 34-year-olds will be studying ideas drawn from these subjects? Probably we cannot. The actual topics of study are not listed, but respondents' occupations that might imply science and technology were "process, plant and machine operatives", "skilled trades", "professional occupations", and "associate professional and technical". Within these categories, however, all subjects are included. Given the wider evidence from elsewhere about interest and engagement with science, one might conjecture that the number studying science or technology would be relatively few.

Only a quarter of those who left school in their mid-teens said that they were learning anything at all. As King, Nomikou, Archer, and Regan (2015) observed: "despite decades of intervention, participation in science still remains strongly patterned, with working-class, women and some minority ethnic groups remaining under-represented in post-16 science" (pp. 2987–2988). The lack of young women leaving formal education without qualifications in science and technology continues to be a matter of particular concern for many countries, for example in Australia (Chubb, 2012; Office of the Chief Scientist, 2016). Yet, as our case-stories will show, there are at least as many women as men who have a need to access helpful science informally, to address a variety of needs in contemporary life. The overall trends of under-representation across socio-economic and gender boundaries seem likely to be replicated with older age groups. In the NIACE (2015) survey, older age groups were very unrepresented among the "learners". This is a matter of concern: "learning for leisure and personal interest is particularly important for older learners, those who are not in employment and those without internet access" (p. 5).

In Australia, but related specifically to science and technology, an interview survey of a representative sample of 800 people from the State of Victoria found people aged 18–44 years were more likely to search for information about science and technology than people aged 45 and over and, at all age levels, men were more likely than women to search for such information (SweeneyResearch, 2011).

In summary, then, and generalising from the NIACE survey, few adults are engaged in learning science unless it is for work or to further their career. The proportion of adults who are likely to be interested in self-initiated learning of science and technology through formal courses, for their own interest, is therefore probably quite small at the moment.

If aims of achieving a scientifically engaged public are to be realised, this must be addressed. There is no doubt, however, that many adults do access scientific information for various reasons that they might not classify as "learning" in terms of the NAICE definition above. This book seeks to illuminate how and why that happens. Our case stories recount the ways in which people set about *finding out* and *understanding* the information they needed for a particular goal, rather than "studying" or "practising" science.

Across any adult population, the degree to which relevant scientific knowledge is held by individuals will vary greatly. Here we are particularly concerned with relevant knowledge that individuals possess and, in particular, how they might harness, extend, or enrich it to meet their own personal purposes. Active learning – for that is what is needed – will only take place if the learner has a conscious and productive notion of what they wish to learn. Further, any person who is involved in supporting that learning must have a view of their role that is commensurate with that notion of learning. Meeting these conditions is made especially difficult because there is no general agreement on what learning by adults involves. In the next section we will provide broad sketches of the major paradigms of learning that are currently thought applicable to adults, pointing to their strengths and weaknesses. We will conclude that section with a statement of the notion of learning that we think best suited to the use of adults who are engaged in self-directed learning.

Models of Learning

As the achievement of "understanding", as such, is so highly valued for personal, cultural, and economic reasons, much effort has been expended in trying to work out what "understanding" involves and how it takes place. Because of the opacity of language and the sheer impossibility of identifying (so far!) exactly what the behaviour called "learning" actually involves at brain level, many models of it have been produced over the years. As the purpose of this book is to see how self-directed learners can be helped in their self-appointed tasks, we will discuss only the broad outlines of four major models of learning by adults and will not become lost in the minutiae of their details. We present them in the order of their increasing relevance to the self-directed learner.

Models Underlying a Classical View of Pedagogy

In the absence, until fairly recently, of systematic enquiry into how adults learn, this model was derived as a parody of what takes place – or is thought to take place – in schools. The emphasis is on teaching rather than on learning. The main tenets of this view are as follows:

- Learning takes place in obedience to the wishes of a social authority (a teacher, an education authority) as represented by "the curriculum".
- That curriculum is heavily shaped by the content of "subjects", for example, physics, chemistry, rather than on specific themes and issues.
- What is learned and when it is learned is determined by a teacher who has control of the whole process.
- The basis of this design is the expectation of "transmission". That is, facts are presented in a predetermined order, the intent being that the learners acquire a "mental copy" of those facts which can be reproduced during assessment.
- A fixed design of learning is used by the teacher (the "lesson"), one which is not flexible to the needs and interests of a particular learner.
- The teaching methods used are limited in scope, variations on the "lecture" being the most commonly used.
- The learner does what is required in order to progress to the next stage of the pre-determined curriculum, rather than having any intrinsic reason for doing so.
- Only a limited set of materials is available to the learner and this defines what is to be learned. Typically, this is a textbook designed around the mandated content of the curriculum.
- Little substantive use is made of the learners' experience, even that which may be relevant to the topic under consideration.

When this set of assumptions is made, the learning that takes place is informed by "behaviourist psychology"; the assumption being that the student knows nothing about the subject in question (here a science) until "given" knowledge by or on behalf of the teacher. Alas, there is evidence that this model is still used to some degree in many schools, despite evidence that many school-age students either do not learn what is required of them when exposed to this approach, or learn distorted versions of what is intended, or just forget it quickly. When this approach has been adopted by, or on behalf of, the self-directed adult learner, the outcomes are also substantially negative. Thus, this content-transmission approach, where still adopted by informal education agencies (for example, by some museums), results in a failure to persist in efforts to learn in many cases (National Research Council, 2009).

Constructivism

Constructivism took root in school science education as a reaction to the evident failure of many students to learn when taught using the classical pedagogic approach described above. While this approach has many roots, a major source is the "humanistic psychology" of Carl Rogers (Rogers, 1969, p. 5). The focus here is directly on the learning that takes place, rather than on the teaching provided. It makes the following assertions:

- Learning involves the whole person, the sum of their cognitions, feelings, beliefs.
- Even when learning is initiated by external events, for example the demands of employment, the "understanding" actually achieved is internal to the learner.
- Effective learning is that which leads to a widespread change of behaviour, attitudes, even beliefs, by the learner.
- The learner evaluates the worth of what is learnt. Only that knowledge which is seen to be of value against some personal criterion is retained.
- This sense of value constitutes the meaning for the individual of what is learnt.

A massive influence exerted on the constructivist approach to learning at its core was due to Jean Piaget (see, for example, Piaget, 1929), who suggested that acquired knowledge was organised into networks, what he termed "schemas". He also suggested that learning involved either: the "assimilation" of new knowledge into existing schemas; or the "accommodation" of that knowledge into existing schemas that are changed to incorporate it; or into new schemas that are created. Piaget asserted that the capabilities of individuals to learn in any or all of these ways underwent qualitative changes as individuals moved from early childhood through adolescence. Alas, he never seems to have tested his "development of capability with age" ideas in respect of adults. We can only assume that the highest level of attainment achieved in adolescence is that eventually obtained by all adults. Despite this, Piaget's notions of the formation and evolution of schemas were adopted by later research workers for use in the context of all learners and have had a considerable impact on adult education.

Thus George Kelly (1955) saw learning entirely in terms of the personal schemas, what he termed the "constructs", of the individuals concerned (see, for example, Pope & Keen, 1981). According to Kelly, there is an infinite variety of ways in which the world-as-experienced can be construed. An individual learns in order to represent and hence to anticipate events in the world as personally experienced. These representations are used to underpin predictions about present and future events that are then used to guide the behaviour of the individual. The model of the world so constructed is continually changed with the intention of producing ever-more successful predictions. For this reason, Kelly's approach is often summarised as "man [*sic*] the scientist". Communication between individuals is only possible when their constructs of reality are substantially alike. Thus, the roles of the teacher are, first, to support the student in developing personal views of the aspect of reality under consideration and, second, to facilitate the evolution of those views towards a more productive view, one that is thought to be of value by the general community. Thus, Kelly's approach suggests that self-directed learning is entirely feasible; indeed it suggests that it is the only authentic approach to learning.

The influence of Kelly on the discussion of adult learning has been eclipsed in recent years by growing attention to Lev Vygotsky's socio-cultural perspective (Vygotsky, 1978). In this perspective, an individual's knowledge of a field of enquiry is manifest in the ability to solve problems within it. The acquisition of understanding takes place via social interactions mediated by language and from the acquisition and use of the specialist symbols of representation used in the field. Learning can result in two ways: first, from a "scaffolding" of ideas concerning the chosen field of inquiry that is erected by someone who is more knowledgeable about the field of inquiry than the learner (a "teacher"); and second, from the joint construction of knowledge by the student and her or his peers. The extent to which learning can take place on any one occasion is defined by the problem-solving capability of the student working alone and that which can be attained by working with a teacher or more knowledgeable peers; this is denoted by the term "zone of proximal development". The use of language is central to Vygotsky's view, for learning initially takes place on the verbal, social, inter-personal plane with knowledge subsequently being internalised by the use of intra-personal language. The inter-personal processes can be monitored by the analysis of the discourse that takes place and then, later, by the subsequent problem-solving capability that is displayed.

One of the major manifestations of constructivism in school science education has been the emphasis, at least in theory, placed on inquiry-based learning. A broad and loosely defined approach, it has generated an extensive research literature (Crawford, 2014). The core idea is that, instead of merely receiving established knowledge from a teacher-as-instructor, students should seek their own answers to questions about natural phenomena. The implementation of the approach goes through four phases: first, the confirmation of the basis of established knowledge; second, the teacher providing the questions and the outline of the processes to be followed in the inquiry; third, the teacher only providing guidance for students' inquiries; fourth, students conduct full enquiries from question-posing through procedures designed and implemented to the communication of the results and the conclusions reached.

Although inquiry-based learning has been universally adopted in national curricula, it is not clear how extensively the inquiry approach is practiced in the classroom, for implementing it faces a number of major challenges. Three of these challenges are that teachers and students will have to modify their roles their roles in the classroom from that of instructor/recipient to that of facilitator/questioner; the size of the curriculum will have to be reduced, for inquiry takes a lot of time; and the style of management of the classroom has to change dramatically. Nevertheless, inquiry-based learning is being implemented to varying degrees in different classrooms. It is therefore likely that, as the years pass, the proportion of the adult populations who have encountered it, in some form or another, will gradually increase. This should mean that they are somewhat more inclined and able to manage their own learning of science.

Although the notion of "constructivism" as a general approach to learning has been subject to a steady stream of criticism over decades (see, for example, Matthews, 1994, pp. 137–161; Osborne, 1996; and many others), the ideas of Vygotsky in particular have been slowly adopted in schools and have gradually permeated formal adult education. The approach is of value both in the context of the structured, taught courses and for the self-directed learner. The general features of it follow.

- Learning involves accommodation and assimilation (see above).
- What is learnt depends on the cultural and educational background of the learner, that is, what they already know.
- While what is learned by the young strongly reflects the ideas of the teacher, a progressive independence of thought is achieved with age.
- To be effective, the teacher must act as a "facilitator", relating what is to be learnt to what the student already knows.
- For learning take place, the student must be mentally active, considering the ideas being presented in relation to existing ideas.
- What is to be learnt is best presented in a context of the use of the ideas involved in it.

However, in the 1970s, before constructivism became established first in school science education and then in adult science education, Malcolm Knowles developed a distinct approach to adult learning which he called "andragogy" that has become firmly established in adult education in general (see, for a recent version of it, Knowles, Holton, & Swanson, 2015).

Andragogy

This approach makes two assumptions: First that all adults learn in ways that are qualitatively different from those used by all school-age students. Second that the great majority of adult learning takes place in formal classes. In more detail, these assumptions are that:

- Adults need to know why they should learn something before they are willing to learn it.
- Adults have control of their own lives and need to feel that this should be recognised by others, for example by teachers.
- Adults have had many more experiences of a wider variety of contexts than those had by school-age students. This range means that they think about things rather differently than when they were young.
- Adults wish to acquire knowledge that will have an impact on their everyday lives.
- Adult learning is problem-oriented, that is, they wish to be able to acquire knowledge that will be of some definite, preferably immediate, use to them.
- In addition to responding to external pressures to learn, for example in order to gain promotions at work, adults also respond to internal pressures to learn, for example in order to feel some increased self-esteem.

(Knowles, Holton, & Swanson, 2015, pp. 63–68)

Central to andragogy is the process of self-directed learning and the most cited definition comes from a guide for self-directed learning by Knowles (1975).

> In its broadest meaning, self-directed learning describes a process in which individuals take the initiative, with or without the help of others, in diagnosing their learning needs, formulating learning goals, identifying human and material resources for learning, choosing and implementing appropriate learning strategies, and evaluating learning outcomes.
>
> *(p. 18)*

The self-direction or independence of the learner means that the role of the teacher is very much that of a guide or facilitator who supports and encourages the learner to develop their capacity to find and use information and solve problems in everyday situations.

Although of great value in advancing adult learning, Knowles' ideas have, over the years, attracted a range of criticisms (Jarvis, 1987; Tennant, 1996). Knowles' original model seemed to assume that learning took place only in formal classes led by a teacher, while much learning actually takes place in other, informal, contexts. In his later work, the "formal class" – "individual learner" distinction is relaxed, but it was assumed that all school-level learning was framed by classical pedagogy. The essence of this assumption is that, as a parody, younger students are incapable of directing their own learning, and students have not had enough experience of the world and its affairs to provide a worthwhile contribution to their own learning. Knowles seemed to assume that students are not able to direct their own learning and thus are not able to engage in problem-based learning. Such assertions about school-level students are no longer universally true, if they ever were. The rapid evolution of the Internet and the social media, coupled with higher levels of prosperity in many countries, have worked against any truth of these assumptions, even if they were ever valid.

There is a continuum in the learning characteristics of "young people" and "older people" that is now not seen to be rigid but is largely influenced by the nature and context of the task being undertaken. Knowles' approach assumed that "education is value-neutral and apolitical", but sociological analysis has challenged this notion (Merriam, Caffarella, & Baumgartner, 2012, pp. 87–89), and also the idea that all adult learning takes place in the same way by different people. Chapter 7 in this book presents an extended example that disproves both of these assertions.

Despite its criticisms, andragogy and its focus on the self-directed learner dominated thinking about adult learning and teaching for several decades. The advent of the Internet in 1991 and consequent increases in the amount, availability, and accessibility of information, together with the increasing need for flexibility in the workplace and organisational structures, have thrown emphasis on the role of the learner to determine what, where, and how learning will occur. Hase and Kenyon (2000) suggested a new educational approach was needed and called it "heutagogy", the concept of learners determining their own learning.

Heutagogy

Heutagogy, according to Hase and Kenyon (2007), "is concerned with learner-centred learning that sees the learner as the major agent in their [*sic*] own learning, which occurs as a result of personal experiences" (p. 112). The approach may be considered to be a logical extension of both pedagogy and andragogy, but rather than developing competence in acquiring knowledge and skills, heutagogy emphasises the capacity to take this knowledge and those skills into unknown situations. It has proved of increasing interest because of a growth in the use of new technologies in distance learning (Blaschke, 2012). Earlier, Hase and Kenyon (2000) had warned that distance education through the provision of teacher-prepared materials may be described as "flexible delivery" but it does not imply flexible learning. These authors argue that:

> a heutagogical approach recognises the need to be flexible in the learning where the teacher provides resources but the learner designs the actual course he or she might take by negotiating the learning. … As teachers we should concern ourselves with developing the learner's capability not just

> embedding discipline based skills and knowledge. We should relinquish any power we deem ourselves to have.
>
> *(p. 4/7)*

Thus the essence of heutagogy is an assertion that all learning is in practice entirely self-directed, with learners only learning about what is of interest and relevance to them. It also involves the development of their capacity to reflect on the processes involved in learning. It has these core ingredients:

- The learner decides what is to be learnt.
- What is being learnt may change, to a greater or lesser degree, during the process of learning.
- The identification of questions to be answered by the learner provides the dynamic of the learning process.
- The learner develops criteria that are to be applied to what is learnt in order to see if that meets the requirements that were set out for, or developed during, the learning.
- Self-reflection – its development and application – is the key process in all the above ingredients.
- The function of any teacher involved is to act as a tutor rather than as an instructor, supporting the learner to address all the above ingredients.

At this point we add a note about autodidactism, a term some authors use to mean "the art of self-education" (see, for example, Stansfield, 2014), which is consistent with the heutogogical view of learning as being entirely self-directed. In his blog, Stansfield (2014) stated that being an autodidact means taking responsibility of one's own education in choosing what to learn and how to learn it. He also pointed out that the rapid increase in accessibility of knowledge via the Internet, for example, makes being an autodidact easier. Solomon (2003) used the term "to describe a range of people who prefer to teach themselves or to pick up knowledge from non-teaching situations, in one way or another" (p. 3). Solomon's edited book contains a mix of stories about various autodidacts in diverse situations and one common aspect was their resistance to being taught, something that was easy to discern in educational situations. Clearly, autodidactism is not an either/or state; any learner can be an autodidact some of the time. In this book, our interest concerns those adults who have a real need to learn, so we turn now to a synthesis of what our exploration of learning has revealed.

The Main Elements of the Self-Directed Learning of Science and Technology

What do the models of learning described above have to say to the adults who, for personal reasons, need to learn more about science and technology? Drawing on the common humanistic threads in constructivism, andragogy, and heutagogy, we suggest that the main challenges that arise as a learner tackles a chosen learning task will be finding ways to do the following:

- Identify the specific knowledge and skills that form part of the chosen sector of science and technology.
- Develop personal criteria for success in respect of the knowledge and skills to be acquired and in respect of the particular reasons for acquiring it.
- Draw on their prior experiences that seem relevant to the chosen task.
- Make direct and immediate use of the knowledge they acquire to solve specific problems.
- Reflect on what was involved in gaining the desired knowledge, and deciding "Have I learned enough, at least for the time being?"

In Chapters 3 to 6 we present case stories of individual adults learning specific aspects of science and technology for personal reasons, pointing out the challenges involved and how they were addressed. We will be testing the points above in the context of our case stories and this will enable us to evaluate the success of self-directed learning as a route to furthering science knowledge and scientific literacy more generally, and the ways in which the self-directed learner may be supported in the summary chapters of this book. First, however, having examined models of learning, we turn to the reasons why adults may want to learn more science and technology.

Theories of Motivation and Self-Determination

If individuals do successfully (in their own terms) learn specific science and technology for personal reasons, they can be said to have acquired motivation to achieve that particular task and to have exercised self-determination by focusing on it. These psychological states – acquiring motivation and exercising self-determination – are inherently abstractions that have been the subject of debate for many years. In this section, we explore the theoretical underpinning of the terms "motivation" and "self-determination" as they relate to how adults pursue their own learning in science and technology.

Why do people do what they do? A person's actions and behaviour are commonly interpreted in terms of their motivation: "To be motivated means *to be moved* to do something" (Ryan & Deci, 2000, p. 54, original emphasis). Prior to 1950, motivation theories were based primarily on the association between a stimulus and response, in which a person's actions were motivated by the expectation of particular outcomes. From the late 1950s, many psychologists took a more cognitive approach, paying attention to the internal processes governing behaviour and recognizing that "out-put" behaviour could be determined by more than an external stimulus. For example, White (1959) proposed the concept of "effectance motivation", an internal kind of motivation that could explain volitional behaviours, like play and exploration, that did not depend on reinforcement for maintenance but on the internal feeling of competence. This is often referred to as intrinsic motivation because the source or stimulus is internal. In contrast, extrinsic motivation has a source or stimulus external to the person. Intrinsic and extrinsic motivation can also be differentiated according to the reasons for action or their outcomes, as follows:

> When intrinsically motivated, people engage in activities that interest them, and they do so freely, with a full sense of volition and without the necessity of material rewards or constraints. Extrinsically motivated behaviours, on the other hand, are instrumental in nature. They are performed not out of interest but because they are believed to be instrumental to some separable consequence.
>
> *(Ryan & Deci, 2000, p. 55)*

Deci and Ryan (1985) recognised the centrality of volition in intrinsic motivation and proposed a theory of "self-determination". They argued that "when people are intrinsically motivated, they experience interest and enjoyment, they feel competent and self-determining, they perceive the locus of causality for their behaviour to be internal, and in some instances they experience flow" (p. 34). This perspective was reinforced by Deci, Vallerand, Pelletier, and Ryan (1991): "Intrinsically motivated behaviours represent the prototype of self-determination – they emanate from the self and are fully endorsed" (p. 328).

Deci and Ryan (1985) argued that self-determination is not only the capacity to choose and to have choices, it is also a human need. "The psychological hallmark of self-determination is flexibility in managing the interaction of oneself and the environment" (p. 38). According to Deci et al. (1991), self-determination theory attends to the human needs for competence (being able to achieve desired outcomes), relatedness (developing satisfying connections with others), and autonomy (self-initiating and self-regulating of one's own actions).

Deci and his colleagues were unequivocal about the importance of volition and choice in human behaviour, and this has important implications for adult learning. Self-determination is related to control. When a person believes that he or she has choice in what is to be learned, then self-determination is the governing principle. However, sometimes people's choice of behaviour is externally motivated by the value attributed to the outcome. Is this also self-determined behaviour? Deci and Ryan (1985) pointed out that "the concept of self-determination permits a more refined and elaborated conception of extrinsic motivation" (p. 35), and suggested that the perception of an internal locus of control "is integral to intrinsically motivated behaviour and is also evident in some extrinsically motivated behaviours" (p. 38). Such behaviour enables an individual to formulate personal values and goals (p. 130).

Deci et al. (1991) and Ryan and Deci (2000) described four levels of extrinsic motivation, according to the extent to which the regulatory process and control become internalised. An external locus of control means that outcomes are externally regulated, so extrinsic rewards and/or punishment are most salient to action. If there is ego involvement and the need for approval from one's self or others, the perceived locus of control is not entirely, but "somewhat" external, and Deci and colleagues referred to this as introjected regulation. When the perceived locus of control becomes "somewhat internal", with a person consciously valuing the activity with self-endorsement of the goals, this is described this as identified regulation. Integrated regulation occurs when "the regulatory process is fully integrated with the individual's coherent sense of self; that is, the identifications are reciprocally assimilated with the individual's other values, needs, and identities" (Deci et al., 1991, p. 330).

These differences, which determine much about motivation and self-direction, are often easy to discern within formal education. How important are they, however, when adults learn informally and individually? This question will be considered when analysing the case stories in this book, and more explicitly in Chapter 10.

Individual Engagement with Science and Technology

If we wish to gain some leverage on the reasons for a person's motivation and self-determination to learn some science and technology, one place to start is to identify personal circumstances that might have triggered such an activity. In general, such circumstances may be urgent, such as a medical issue, or perhaps less urgent, such as the need for progress at work, or a single episode requiring new knowledge. As a person engages in a particular type of work, for example, it is sometimes the case that new opportunities present themselves. Often building to some extent on existing experience, such an opportunity requires a "leap into the dark" in which substantial new knowledge of science and technology is mandated. The circumstances may also be ongoing – exploration to develop a hobby or a general interest, for example. Such interest could also be concerned with a wider issue, such as climate change or nuclear waste.

In order to gain some insight into our broad questions, we asked a range of people, chosen for their diversity, if they had recently engaged in learning science or technology for one of the types of reason that we had identified. We then invited the "opportunity sample" so identified to write us 1,000 plus words on their experience that was shaped in response to the following five questions:

1. How did the need to know the new science and technology come about?
2. In what ways was that knowledge new to you?
3. Where did you go to find the knowledge? (Whom did you approach? What resources did you search for?)

4. What particular aspects of those places, people, resources, were especially helpful to you?
5. What, if anything, could have made your task easier?

Each case story was discussed with the appropriate self-directed learner in order to clarify any ambiguities. The final versions are fully cited, with the agreement of the authors, in Chapters 3, 4, 5, and 6. We have attached a commentary to each case story – again with the agreement of its author – in order to draw out issues salient to adult learning of science and technology in general. In subsequent chapters, we overview the kinds of resources accessed and how they were used successfully, or not, by our storytellers, and then draw together a framework describing how best another person, not necessarily a teacher, might assist the self-directed learner in their learning journey.

References

Blaschke, L. M. (2012). Heutagogy and lifelong learning: A review of heutagogical practice and self-determined learning. *The International Review of Research in Open and Distance Learning*, 13(1), 56–71.

Chubb, I. (2012). *Mathematics, engineering and science in the National Interest.* Canberra, Australia. Retrieved from http://www.chiefscientist.gov.au/wp-content/uploads/Office-of-the-Chief-Scientist-MES-Report-8-May-2012.pdf

Crawford, B. (2014). From inquiry to scientific practices in the school curriculum. In N. G. Lederman & S. K. Abell (Eds.), *Handbook of research in science education* (Vol. 2, pp. 515–544). New York: Routledge.

Deci, E., & Ryan, R. (1985). *Intrinsic motivation and self-determination in human behaviour.* New York: Plenum Press.

Deci, E., Vallerand, R., Pelletier, L., & Ryan, R. (1991). Motivation and education: The self-determination perspective. *Educational Psychologist*, 26(3), 325–346.

Fensham, P. (2008). *Science education policy-making: Eleven emerging issues.* Paris, France: OECD. Retrieved from http://unesdoc.unesco.org/images/0015/001567/156700E.pdf

Fensham, P., & Harlen, W. (1999). School science and public understanding of science. *International Journal of Science Education*, 21(7), 755–763.

Hase, S., & Kenyon, C. (2000). From andragogy to heutagogy. In *UltiBase Articles.* Retrieved from http://ultibase.rmit.edu.au/Articles/dec00/hase2.htm

Hase, S., & Kenyon, C. (2007). Heutagogy: A child of complexity theory. *Complicity: An International Journal of Complexity and Education*, 4(1), 111–118.

Jarvis, P. (1987). Malcolm Knowles. In P. Jarvis (Ed.), *Twentieth century thinkers in adult education* (pp. 144–159). London: Croom Helm.

Kelly, G. (1955). *The psychology of personal constructs* (Vol. 1). New York: W.W. Norton.

King, H., Nomikou, E., Archer, L., & Regan, E. (2015). Teachers' understanding and operationalisation of "science capital". *International Journal of Science Education*, 37(18), 2987–3014.

Knowles, M. (1975). *Self-directed learning: A guide for learners and teachers.* Englewood Cliffs, NJ: Cambridge Adult Education.

Knowles, M., Holton, E. F., III, & Swanson, R. A. (2015). *The adult learner* (8th ed.). New York: Routledge.

Matthews, M. (1994). *Science teaching: The role of history and philosophy of science*. London: Routledge.

Merriam, S., Caffarella, R., & Baumgartner, L. (2012). *CourseSmart-Learning in adulthood: A comprehensive guide*. San Francisco, CA: Jossey Bass.

National Research Council. (2009). *Learning science in informal environments: People, places, and pursuits*. Washington, DC: The National Academies Press.

NIACE. (2015). *2015 NIACE Adult Participation in Learning survey*. Retrieved from http://www.learningandwork.org.uk/resources/2015-adult-participation-in-learning-survey

Office of the Chief Scientist. (2016). *Australia's STEM workforce: Science, technology, engineering and mathematics*. Canberra, Australia: Australian Government. Retrieved from http://www.chiefscientist.gov.au/2016/03/report-australias-stem-workforce/

Osborne, J. (1996). Beyond constructivism. *Science Education*, 80(1), 53–82.

Piaget, J. (1929). *The child's conception of the world*. London: Routledge and Kegan Paul.

Pope, M., & Keen, T. (1981). *Personal construct psychology and education*. London: Academic Press.

Rogers, C. (1969). *Freedom to learn*. Columbus, OH: Merrill.

Ryan, R., & Deci, E. (2000). Intrinsic and extrinsic motivations: Classic definitions and new directions. *Contemporary Educational Psychology*, 25, 54–67.

Solomon, J. (Ed.). (2003). *The passion to learn: An inquiry into autodidactism*. London: RoutledgeFalmer.

Stansfield, S. (2014). Autodidactism: The art of self-education. [web log comment]. Retrieved from https://peregrinepoise.com/autodidactism-the-art-of-self-education/

SweeneyResearch. (2011). *Community interest and engagement with science and technology in Victoria*. Melbourne, Australia.

Tennant, M. (1996). *Psychology and adult learning*. London: Routledge.

Vygotsky, L. (1978). *Mind in society: The development of higher psychological processes*. Cambridge, MA: Harvard University Press.

White, R. (1959). Motivation reconsidered. *Psychological Review*, 66, 297–333.

3

LEARNING TO DEAL WITH MEDICAL ISSUES

Maintaining one's health for as long as possible is perhaps the issue that causes a person to most actively seek scientific knowledge of human biology and its associated technologies. This is a vast field and the health issues that people need to deal with vary from small injuries to life-threatening illnesses. According to the nature of the complaint, people pursue many different paths to find out more about their condition and, if possible, remedy it.

Seeking Solutions to Health Problems

We have obtained case stories from three people for whom the search was of increasing importance and urgency. These are: Ana's story, which concerns her first pregnancy; Penny's story, which concerns episodes of ill-health, the cause of which was difficult to diagnose; and Mary's story, which is focused on Laura, her daughter who was born with Down Syndrome. This range of stories throws some light on how individuals seek medical knowledge with varying degrees of urgency.

Ana's Story: First Pregnancy

When Ana and Joao decided to have a baby when she was in her late-30s, her only formal knowledge of what pregnancy involved was derived from a course on human biology taken in her high school years. She had however studied the physical sciences extensively since that time, holding a MSc in Science Education (from a Portuguese university) and a PhD (from a United Kingdom university). Her knowledge of how to study a science, derived from those experiences, proved invaluable to her.

DOI: 10.4324/9780429434303-3

After realising that I was pregnant with my first child, I was interested to know in more detail about three issues. First, about the development from the embryo to the baby and the link to my psychological and physical states, in particular how pregnancy would affect my moods and when would I start feeling tired. Second, the kind of medical examinations to which I would be submitted. For example, before being pregnant I knew about echo sounding (2D and 3D) and amniocentesis. I also heard (from some of my friends) about some non-invasive techniques (like blood markers that provide hints about genetic abnormalities). Third, the type of body care I would need to keep in mind. This latter was one of the themes about which I knew more before being pregnant. Not that I needed to search for information because my friends were very worried about having a similar body shape before and after pregnancy but without putting the nutrition needs of the baby at risk. I already knew about: the best lotions to prevent stretch marks during pregnancy; the best clothes to wear after labour; the need to walk often; to use elastic stockings; to drink a huge amount of water, only drinking one cup of coffee per day; to avoid chocolates and cookies during pregnancy. However, I did not know whether or not I could keep going swimming and until when; whether or not I could dye my hair; what kind of soap should I use, etc.

I already knew some facts from secondary biology education (11th grade, mainly about the fertile period of a women, how fertilisation occurs, and the overall processes of transformation of an embryo into a baby). There was a referendum in Portugal about the legalisation of abortion. During this period, there were several debates on the TV and in the newspapers about that theme and arguments pro or against the legalisation of abortion which were often supported by scientific facts about embryology. Many scientists and members of ethics-concerned communities presented their expertise which was then used against or in favour of abortion. I remember some interesting debates from this period, such as: when does foetal pain occur (a theme still under discussion)? How does a stressed mother affect the immune system of the new life due to the accumulation of high quantities of some hormones? What is the percentage of success of implants of an embryo? I remember long and earnest discussions among groups of friends about the referendum and about which way to vote. Many myths about science and technology were put forward in these discussions.

Apart from these episodes, I do not remember about engaging in discussions about the issue of pregnancy as such. Sometimes there were items on the TV news about assisted reproduction. I did find out about the problems that were met during the pregnancy of a friend, but I did not devote much time in finding out specific information. However, it gradually became clear that I needed to know much more in general about pregnancy.

During my first meeting with my doctor, I asked him for a book that could answer some of my questions. He suggested some introductory books on embryology and he was certainly willing to answer all my questions about pregnancy. I borrowed a book from him although, as it was an introduction to embryology, it was very technical and difficult to follow. I found other books on the market but they were very simple and provided a very romantic description of pregnancy. There were also some magazines covering a range of themes from pregnancy to "toddlerdom" (like: "parents & sons"; "problem babies and super-babies"), I bought one of each of them but found that the information in each exemplar was very simple and did not really explain anything. Since I couldn't find a book that could systematically and rapidly address my questions, and because I already knew by then something about pregnancy, I decided to take a different approach. More or less at the beginning of each month of pregnancy, I revised the weekly development of the new life and the nature of transformation I would expect to take place in me. Other issues, such as food, medical issues, I would attend to when they appeared.

The stages of development of the new life and the mother's transformation during that process was very easy to find out about in any book dealing with pregnancy. An encyclopaedia that I have proved very useful. Concerning other matters I had to deal with, two were:

Diabetes – At the beginning of my pregnancy, my levels of glucoses were at the maximum recommended. According to the doctor, the situation was likely to become worse during the later months of pregnancy and, in the limit, I would have to inject insulin. Since I was not then suffering with gestational diabetes, he did not send me to a nutritionist but we agreed that I would not eat summer fruits (e.g. melon) or white bread. At that time, although I knew that diabetes was very serious, I was not completely aware of the extent of the problem. I decided to have a look at the websites of the United Kingdom Department of Health and the abstracts of some scientific papers obtained through the website "Science Direct". I soon realised that mothers could become obese after pregnancy and that I could develop diabetes Type 2. Neonatal diabetes is unlikely but the child can develop other diseases such as obesity, intolerance to glucose, or cardiovascular disease. So I decided to change my diet, reading all the labels in order to choose, for example, the best type of yogurt or bread. I also decided to have a schedule for exercise (swimming 3 times per week). I measured my blood glucose level in a pharmacy. This was not very accurate but it gave me some hints and made me more confident before the next serious blood test.

Amniocentesis – Before doing amniocentesis, I had to sign an informed consent form. Thus, I knew something about the procedure before having the intervention, that is it would involve echography (echo location), I knew the amount of amniotic liquid they would remove, knew something of the safety procedure to be followed. However, it was only on the day involved that I had a real perception of how invasive the procedure was. This was so because, after the intervention, I asked the doctor to show me the needle. I was also able to see the colour of the amniotic liquid and the doctor told me about its composition. Spontaneously, the doctor decided to tell me how the technical procedure evolved. It was an interesting talk.

During pregnancy, I attended a course at the Portuguese National Health Centre about "preparation for child-birth" (for example, how to breathe during child-birth) and how to take care of the new-born (for example, how to change diapers, how to breastfeed the baby). The course was given by a nurse specialized in obstetrics. The public, who attended the sessions, was diverse (some without schooling, some teenagers who attended the course with their mothers). The course was of no use to me because my daughter was born by Caesarean Section and because all the other issues were so basic that I already knew about them. Nevertheless, the discussions that I had with the nurse at the end of the sessions were important. I just went to the course in order to have an opportunity to talk every week with the nurse at the end of the session and to expose my doubts. That was very useful. I remember that we talked about what could go wrong during child-birth, about the need to monitor the baby activity in the belly during the last months of pregnancy, about the fact that some nurses have different points of view about feeding the new born (for example, some believe that they only need the breast, others believe that we also need to provide a supplement powder milk and allow the baby to decide whether he/she wants it); about the need to have a timetable for the production of the urine of the new born during the first days in order to see whether or not the kidneys were working; to allow the new born to have a hearing test; to sing the same songs during and after child-birth. The problem is that, despite the amount of information that can be collected before child-birth, we are never prepared enough for the actual event. It is impossible to anticipate all the issues that one will face. Only the experience itself gave me a feeling of control. My daughter cried day and night, she was unable to digest breast milk properly and she didn't "like" the powdered milk. The doctor helped solving the problem: fennel tea!

My learning process was not hard. All the doctors and nurses with whom I interacted were very helpful in answering my questions. Some booklets with information about diabetes, iron deficiency, amniocentesis, and blood tests would have been useful.

Ana's case story is both typical and atypical. Typical in that it addresses the challenges that many, if not most, women face during and after pregnancy. She became pregnant in her late 30s, as do many other women, after having met many other, very different, challenges derived from a busy professional life. Atypical in that her degrees in science education had evidently introduced her to the issues involved in self-initiated learning. She did face some novel challenges: Not only was the knowledge that she needed to be derived largely from the biological sciences, while her existing knowledge base was firmly in the physical sciences, but also that knowledge was to be directly applicable to her as a person rather than merely being something that she could recall.

Nevertheless, she seems to have actively considered the types of issues that she would face, drawing on the experience of her friends. Her immediate concerns were focused on the physiological and psychological changes that would take place in her during the pregnancy and the implications of the medical procedures that she would undergo. Her responses to these can be summed up as a desire to physically "return to normal" once the baby was delivered. She was aware that the rudimentary biological science knowledge derived from her schooldays was an inadequate basis for producing the explanations of the types she sought, having been made aware of these deficiencies by recent discussions with her friends.

Once her pregnancy had been confirmed, she turned to her general practitioner for information and advice. He seems to have attempted to be helpful, but the book he lent her was too technical and did not address her immediate questions. When she looked in bookshops, she found magazines and books that took either a too romantic view or a too simple approach to explanation. It is interesting that an encyclopaedia at home proved a useful basic source of the information that she sought and gave her ideas about the issues that she would face.

On the basis of that knowledge, she decided to keep a regular monthly record of the changes that she could expect during and after the pregnancy. One major concern was the possible onset of diabetes, which she addressed through a review of her nutrition and the adoption of a regime of regular exercise. She derived useful knowledge from a United Kingdom government database that she knew of from her time as a research student in England. She then found discussions with a nurse attached to the local antenatal unit to be especially valuable in respect of practical, everyday questions about handling the baby and in dealing with the nutritional problems that might arise.

The initial help by her doctor, augmented by access to an encyclopaedia, provided the basis on which both to seek advanced information from the Internet and, most importantly, to have discussions of a very practical nature with an antenatal nurse. Her suggestion of leaflets addressing several major issues is surprising: one would have assumed that they would already be widely available, given that pregnancy is a universal phenomenon.

Penny's Story: An "Invisible Disability"

Penny trained as a high school mathematics teacher and now she has retired, she works as a volunteer teacher's aide, assisting young children with their reading. At no stage in her school or higher education did she take a formal course in human biology, although she had learnt quite a lot while supporting her husband, who has a heart condition requiring a pace-maker. She faced an urgent need to learn when she unexpectedly found herself unwell. Her story is based on a series of emails to a friend so it has a diary format that reveals Penny's contemporary reflections on her condition.

> *Episode One, October*
>
> *At the end of July, I leapt out of bed one morning, a bit later than usual because I hadn't slept particularly well. I'd had a fasting day the day before so I was hungry and raced to the kitchen to start preparing breakfast. Suddenly my head spun and I felt nauseous. I leaned on the bench for a few seconds and it passed so I continued my food prep but then it came back with a vengeance. I decided I was ill and had better go back to bed and the next thing I observed was that my cheek was pressed to the kitchen floor. I thought I might have had a stroke but I tried all my limbs and they seemed normal. The next thing to test was speech, so I shouted to Philip, my husband, that I*

was ill and needed help. He'd heard the crash but thought I'd knocked over a chair. He helped me back to the bedroom where I blacked out again and fell across the bed. We decided this needed investigation and it seemed best to call an ambulance. They took me to the local hospital emergency ward where I was admitted for 24-hour observation. The paramedics measured fairly high BP (blood pressure) and did an ECG (electrocardiogram) which showed very variable gaps between the "spikes" on the trace, apparently suggesting atrial fibrillation. That printout was lost somewhere so the doctors only had hearsay evidence. My BP continued to rise briefly then gradually subsided. At one stage my systolic pressure was 144; when they got me to stand up it dropped to 118. Apparently, a drop of more than 20 is classed as postural hypotension and is common among older patients on hypertension drugs, as I was.

In due course I was told my liver and kidney functions were fine and I hadn't had a heart attack so that was all reassuring. I sat on a bed for 24 hours with various things being monitored, including blood pressure measured every hour. Although there was no difference in my activity, or lack thereof, the systolic BP varied between below 120 and over 140 rather randomly. Periodic ECGs didn't cause any concern until the next morning, when the nurses frowned, moved the sensors and tried again. When I asked if there was anything wrong they said, "No, it's fine." On this occasion I was apparently lacking "P-waves" in the trace. They didn't tell me what that meant, but I inferred that it must be connected with atrial fibrillation because they eventually discharged me with instructions to get a Holter monitor (a mobile ECG device) and wear it for three days to detect heart abnormalities.

It took weeks to get the Holter monitor but apparently it didn't show anything unusual for someone of my age. The net result was that the medical staff felt they'd ruled out all sinister causes. I concluded that leaping out of bed and rushing about after reduced food intake had caused a sudden drop in BP and I'd fainted and although none of them has actually agreed with my diagnosis they didn't deny it. They asked whether I'd felt as though I was going to faint but since I'd never fainted before I had no idea how it felt. Now I know.

While I was waiting for the Holter monitor to become available a friend told me that he'd had a similar fainting experience, wore a Holter monitor for 24 hours, returned it and went back to work. In due course (it takes a while to process these things) he got a phone call at work telling him to come to hospital immediately and "don't drive!". He was fitted with a pace-maker that day. It was all rather nerve-wracking waiting for a diagnosis and I observed that the more you think about how you feel the worse you feel. For a while I was asking the lifeguards at the local pool to keep an eye on me when I was swimming and feeling a bit uneasy when I was out on an orienteering course. I felt fairly normal then and now although I think I do detect a bit of instability in my balance. I take a bit more care about things like descending stairs and I generally try to be a bit less of a bull-at-a-gate than comes naturally.

Reflecting back, I really believe that my fainting was due to a drop in BP when I jumped up and rushed about while being low on "fuel". The root cause would be the hypertension medication I was then taking. Nobody in the medical field seemed to be the least bit concerned about my BP fluctuations. I reduced my hypertension tablet dose by half after that event and when I check it's usually less than 130/70. I suspect I could cut them out altogether without doing any harm. I do consult my GP but usually I tell her what I'm planning to do rather than asking her advice. I must have a rather bullying demeanour because people seldom argue with me. The GP says things like "we don't want it to get too high but then we don't want it to get too low either". I tried an online search, but I wasn't able to find a clear definition of what is too high or too low. By the old rule of 100+age/100 I've never had high BP except momentarily when under stress. By the current guidelines of 120/80, it's often high but nobody seems to suggest treatment for anything under 140/90. It's all very confusing. A recent "Catalyst" program on TV looked at how illness is constantly redefined so more people need treatment. Mental health seems particularly susceptible to disruption because it's so subjective. These days anyone who is a bit unhappy, often with excellent reasons, is classed as clinically depressed and often medicated with drugs which can have terrible side-effects.

Episode Two, December

I went to school as usual to hear some children reading, but found I couldn't remember the names of the children – kids I knew most of in their kindergarten year and have been seeing twice a week all that year. It was quite disconcerting and I felt inclined to leave but I could read and talk normally so I stayed the usual time. If I'd taken my phone I might have called my husband to come and fetch me. Nevertheless, I decided not to go home via the usually busy major highway so I turned away from it. I should have turned towards it then turned off on a particular side street. I've made that mistake before but more disturbing was that I couldn't remember the name of the street I needed to turn into and couldn't recognise the name when I saw it although we drive along it several times a week. I found my way home then took the neighbour's dogs for a walk. I couldn't remember how to turn on my MP3 player which I always use when walking without human company. Next was coffee time and I'd forgotten how to work the microwave. I achieved what I needed to in a clumsy fashion; I'd completely forgotten that you have to turn a knob to set the time. I think I was a bit unsure what year it was. Nobody asked me the name of the Prime Minister! During the morning I completed the newspaper cryptic crossword, with a little help from my husband. He often has a go if I abandon it. When I commented on some of the words he'd filled in he told me I'd done those, so I'd lost some old memories and failed to register some new ones. My memory gradually returned. I can now picture the Year 1 children and assign names to them except for one girl whose name is an American state or city, I think containing a "k", like Dakota. Philip keeps testing me, but I seem to be completely back to normal. I can't help wondering if this is all a sign of Alzheimer's. Most sufferers seem to be quite lucid much of the time in the early stages. I'm seeing my GP next week so I'll discuss it with her. I hope it was just a temporary aberration like some events I've had a couple of times long ago when I couldn't talk coherently. They passed with a little rest.

My visit to the GP resulted in an appointment to have scans at the local radiological centre, which I did. My scans went to the doctors' surgery but my doctor was away so I'm seeing her this week. I called in to the radiology clinic and asked for copies of the reports but I'm not much wiser. The report read "The carotid ultrasound indicated irregular calcific plaque in the right lobe and smooth focal calcific plaque in the left. This has resulted in 16–49% stenosis" (whatever that means). "The brain MRI showed small focus of remote lacunar infarction posteromedially within the left cerebellar hemisphere. No additional abnormality demonstrated. No cause for patient's recent memory loss and confusion is shown." I realise I could look up these terms on Google, but it's easier to get a context from my doctor. The reports give me things to question when I see her so I think they're worth having. I have no experience or knowledge of the sorts of terms used. Definitions such as those I found on the Internet usually involve more terms I don't fully understand and I'm not certain how they apply to my particular case.

Episode Three, January

After saying I wouldn't bother to look up the terms used in my scan reports, I did Google the terms online and the "lacuna infarction" seemed to be associated with strokes, though it wasn't clear whether it indicated I'd actually had a stroke. I didn't seem to have had any of the symptoms they listed except the slight sense of imbalance I mentioned before. The stenoses in the carotids are generally only acted upon if blockage exceeds 70% although the online information did describe surgery to scrape out the plaque or angioplasty to flatten it if it exceeds 50%.

I had my GP appointment and she confirmed that the brain scan indicated a small stroke. We're guessing that's what happened when I passed out in July but there's really no way to know. The GP suggested I talk to a neurologist because she's not very knowledgeable about it, so I have a referral. She thought he might be able to persuade me to go back on statins (the medication I had given up, which was supposedly useful in lowering cholesterol levels) but I assured her he couldn't. Philip does lots of online research on statins and it all seems to confirm that they have small benefits for anyone and are more likely to harm women, even if they haven't already had nasty side-effects. The GP also mentioned warfarin and I said I was very antagonistic to that. Apart from the ill effects like bruising, bleeding and endless blood-testing, according to online resources, there are all sorts of weird dietary restrictions like "you can eat cauliflower but not broccoli" that I'd find very onerous. I'm happy to take low dose aspirin. Treating hypertension was strongly recommended so I agreed to try a different medication from the one I had previously, which didn't appear to affect my blood pressure. I'm fairly convinced that genetics greatly outweigh any lifestyle factors and I can't give up smoking or drinking since I do neither, get rid of extra weight or start exercising, because I am not overweight and get plenty of exercise through orienteering.

Meanwhile it's a bit disconcerting to think I could have another stroke any time. I can't quite "get my head around" what the risks are. Perhaps the neurologist will make it clearer. It seems possible that the amnesia I suffered briefly a couple of weeks ago was another mini-stroke. Perhaps I'll find out if that's likely but most attempts at diagnoses of these physiological anomalies seem to be very vague and inconclusive.

Most of the stuff I read on-line was about regaining mobility and restoring other faculties that have been damaged by strokes. There was lots of emphasis on acting quickly to minimise the damage but it's too late for that. I guess we'll be prepared next time. When my mother had a couple of blackouts in her 70s I thought they may have been mini-strokes but they were never diagnosed so it's mere speculation based on very poor knowledge of physiology. Smart people studied mathematics, physics, and chemistry when I was at school, which have turned out not to be very helpful for understanding my health! Anyway, in the meantime I seem to be functioning perfectly well (perhaps imperfectly but well enough for practical purposes).

Episode Four, March

Well, I visited the neurologist this morning but I don't feel much wiser. He asked lots of questions and did lots of listening and writing notes but not much talking. The main points I got were:

Strokes don't make you faint. The times I passed out in July were almost certainly of cardiac origin, probably the atrial fibrillation which showed up on the ECG the paramedics did and from which the printout was lost. He thinks I probably need to wear a Holter Monitor for 3 weeks to show up an intermittent problem – horrible thought. I'd need to have further "events" before I was willing to go down that path. He didn't seem very impressed by my diagnosis of postural hypotension, based on my often experiencing some dizziness on standing up suddenly and knowing from previous online exploration of definitions of high and low blood pressure that it could be a result of low blood pressure. Since my drop in BP on standing the day I was in hospital satisfied the definition of postural hypotension it seemed a likely explanation for the acute dizziness which preceded my fainting.

According to the neurologist, I have "Migraine equivalent" with symptoms such as visual impairment (bright spots), speech impairment (on a couple of occasions I have been unable to speak coherently) and memory lapses, as on the morning I forgot so many things. I'm lucky I don't get the terrible headaches of regular migraine. He is going to tell my GP of a medication that I think he said controls blood pressure and treats migraine, definitely the latter. He has referred me for an EEG (electro-encephalograph). I'm not sure what that's looking for and he didn't say.

He tested me in all sorts of ways, including a sustained assault with his reflex hammer but gave no feedback. As I was leaving he conceded that all my responses were normal. I said I needed to know what to worry about and he said "Nothing".

He didn't mention statins or anti-coagulants except to ask if I take aspirin and indicate that it doesn't matter whether I continue to do so. He seemed totally unconcerned about the stroke and focussed on the cardiac symptoms. I should keep an eye on the carotid blockage in case it gets worse but he doesn't think it would be causing any symptoms yet.

He was nicely scientific in his questions; he snarled when I said "not long" in answer to a question about the bright spots in my vision. We agreed that it's difficult to express things in ways others understand. When he asked how big my hypertension pills were, I started describing them, then saw from his face that was silly but he hadn't asked the dose. I didn't know it anyway. So, I'm having the EEG on Friday then have to go and discuss it all with my GP.

Episode Five, April

The EEG did not show anything, as far as I can tell. I persuaded the radiological clinic to give me a copy of the results, but it wasn't informative. The unknown terms I looked up online were explained by other unknown terms, making it all rather unhelpful. I asked a friend who used to be a radiologist why an EEG would be ordered and he said it usually was an action taken when the medical profession didn't know what they were looking for. My GP wasn't concerned about it. She scanned the neurologist's very long report and agreed with the advice to monitor my carotid arteries annually. When I asked about it, she prescribed the new drug for hypertension which is not normally used here but is the drug of choice for migraine in Scandinavia. I've been using that for about a month now and my blood pressure seems to be a bit lower but I don't notice any difference in the incidence of the things I associate with migraine, like patchy vision.

In the end, I really don't know what caused my various symptoms. The complexity of medical terminology makes it difficult to sort through information and opinions, and how do you know if you are using the right terms to describe your own symptoms? I have had several medical opinions, and I have my own opinions, but don't have any way to test whether any are right or not. It seems there was a slight stroke, and I assume lots of people have "silent" strokes and are never any the wiser but it implies a greater risk of future strokes. I am sure that aging is part of my problems; for example, I am a bit unsteady on my feet and use a handrail on steep steps that I would have bounded down a decade ago. Meanwhile, I will continue orienteering and other outdoor exercise, like long walks with the neighbour's dogs, bike riding, and enjoying the nice weather.

Penny seems to be suffering from an "invisible disability", which, as explained by the Invisible Disability Association (IDA), "refers to symptoms such as debilitating pain, fatigue, dizziness, cognitive dysfunctions, brain injuries, learning difficulties and mental health disorders, as well as hearing and vision disorders" (IDA, 2016). The problems with such disabilities are that their symptoms are not immediately obvious to a third party (such as a doctor) and are somewhat transient in occurrence. As a consequence, Penny was unable to get a clear diagnosis from her medical professionals, although the latter evidently tried very hard to be helpful, and she had to rely to a considerable extent on her own initiative for diagnosis and treatment. This suggests that, in the terms of Deci, Vallerand, Pelletier, and Ryan (1991), her motivation to act was very substantially self-initiated, with some degree of external regulation being adopted when she found her doctor's advice to be both substantial and effective. To a considerable extent, the progress she made was helped by her experience in helping her husband, who had a heart condition. This had given her confidence in searching the Internet for information. Her major problems were with the meaning of specialist medical terms and their conveyance by the medical authorities.

The issue of language haunts doctor–patient relations. In a general review of the field, Fong (2010) points to those problems that originate from the doctor: a general lack of training in communication skills; the adoption of avoidance behaviour by the doctor resulting from an uncertainty over what it is best to do or say when the diagnosis is unclear; the implicit threat of legal action by the patient if incorrect advice is given. Other commentators, such as Boodman (2011) and Knapton (2014), are somewhat more sympathetic to the patient's point of view, seeing the issue of poor communication as a general lack of "health literacy" on the part of the lay (i.e. non-medical-professional) public. While some components of health literacy – a knowledge of what controls health – have been identified, for example by Graham (2008), the need for an education that leads to health literacy to be included in the school science curriculum has been made forcefully by Roth (2014) on the basis of his own experience.

Mary's Story: A Lifelong Challenge

Mary went to an all-girls school in the 1960s where the only science subject taught was biology, which she enjoyed. After "O" Levels in Suffolk, she changed schools to do "A" Levels at a school in London and studied English and History before then carrying on to secretarial college. Laura was her second child, the first having been born without Down Syndrome.

> *When my daughter Laura was born in Christchurch, New Zealand, she had an enlarged stomach, so that the midwife immediately contacted the on-site paediatrician, who turned up after a long wait. He and his Registrar were checking Laura and making comments in low tones. He then came up to me and advised me that she had Down Syndrome and was very unwell. She was put into an incubator as she was not breathing properly. It turned out that the swollen stomach was due to a leukemoid reaction to excessive white blood cells which was causing her liver and spleen to expand. They thought she had leukaemia but fortunately, after blood tests, found that this was not the case. She was also given an overall scan for any internal cancers. The doctor advised me there were no "masses". I had to ask what that meant. He told me it is a medical term for tumours. She was in an incubator for almost a month. Her other medical problem was hypothyroidism, which I was advised meant taking tablets for the rest of her life. The white blood cell problem, for which there was no medication, resolved itself with time and her stomach slowly returned to normal.*
>
> *I was referred to another paediatrician who told me a little about Down Syndrome. He said that she would be slow to learn and that people with Down Syndrome vary in their abilities, therefore it was unknown at that stage what she might achieve. I did not feel well informed. While in the hospital I was visited by a counsellor. She put me onto the Early Intervention Program, to help Laura with her very low muscle tone, to*

improve her fine motor skills and to help with her cognitive development. The social worker also referred me to another mother to be my "buddy". She had a 4-year-old boy with Down Syndrome. She was helpful and positive and said to be patient as Laura would be slower to progress than other babies, but she would reach her milestones in the end. Another mother from our church had a 26-year-old daughter with Down Syndrome. She told me that her daughter, although slower to understand, and to do things, was leading a very busy life as an adult and she was a popular person in the community. She also advised that life would never be dull, which is very true, as Laura has turned out to have a great sense of humour. Another mother whom I met through a friend, had a son with a different form of intellectual disability. She was negative and said there were permanent hurdles and barriers to cross, although she added that she knew she should not say that to someone with a new disabled baby. That depressed me.

When I went to the United Kingdom for a visit I went into the United Kingdom Down's [sic] *Syndrome Association and found there were many up-to-date books on the subject. I was so excited I bought about 10 books and carried a heavy suitcase onto the plane. The information could be confronting with all the medical, social and educational problems, but it armed me for the future and gave me more confidence to have the knowledge to deal with problems when they arose.*

We came to Canberra in the Australian Capital Territory (ACT) when Laura was six. I was more confident with her by then. I joined the ACT Down Syndrome Association as the Secretary, which meant attending the monthly meetings and taking the minutes, and also participating in discussions, mainly about social issues. I found that Canberra was behind New Zealand in terms of educational issues. In New Zealand most children with Down Syndrome were in the mainstream schools with support. I battled with trying to get Laura educated in the ACT mainstream system. The teachers and the education department kept pointing towards units and special schools, yet at the Early Intervention Centre that I had attended in Christchurch, the books I had read, and the speakers at conferences, all advised strongly that mainstreaming with aide support was the only way to go, as integrating with ordinary children would help with her development, her language and her behaviour. She attended mainstream, then went into a unit in a different school, then into a special school and then back to mainstreaming for her senior years. I believe that mainstreaming was the best in terms of helping her to progress socially.

Laura has a brother, Andrew, who is five years older. In their younger days he played with her, but they are now both in their twenties, and lead separate lives.

Given the great impact that the birth of a child with Down Syndrome has on the lives of the parents, it is surprising that, while the children themselves have been studied, for example by Cunningham and Glenn (2004) into the extent to which those with Down Syndrome are aware of their condition, there have been few studies from the parents' (or other carers') point of view. An early and comprehensive exception is that by Layton, Jenkins, Macgill, and Davey (1993), who summarised all that was then known about Down Syndrome and its implications. This was written after the discovery of the genetic origins of the Syndrome (in 1959) and at about the time that Laura was born, which means that Mary could have had access to much of what is known about the condition. However, this would have been of little use to Mary (had she known of it), for it was written in an obscure publication and in the specialist language of science education.

At the time that Laura was born, antenatal diagnosis of the condition was not routine for mothers-to-be of Mary's age. The immediate and reluctant response of the medical services to Laura's condition on her birth suggest an underlying belief that such children would have short lives of considerable to great disability, as has been widely reported, for example by Gordon (2015). Attitudes at that time were changing, with a growing perception of the developmental potential of children born with Down Syndrome (Stratford, 1994). This trend has continued to today, although the availability of precise antenatal diagnosis and the legalisation of abortion means that, for example, in the United Kingdom at present about 90% such pregnancies are terminated (Gordon, 2015).

Mary did get advice from the medical services, but got most help from other parents and her own initiatives in acquiring relevant books. One major issue, as in both Ana's and Penny's cases, was the complex language used by the medical sector in general. As Lambert and Rose (1996) point out, in discussing another genetic disorder, it is very difficult for the layperson (meaning: a parent) to understand the science and its practical implications.

In many ways, access to information about Down Syndrome (and other genetic disorders) has become much more readily available in recent years, most notably via the Internet. For example, the wide range of information and services now offered by the Down Syndrome Association on the Internet are couched in everyday language. However, much remains to be done in terms of communication. A comprehensive account of what can be attempted and achieved with those who have Down Syndrome throughout their lives, and which would have been very useful to carers, was couched in "high science" language and hence was somewhat inaccessible (Bull, 2016).

Commentary on the Three Case Stories

In all three stories, the person concerned is exhibiting what Deci et al. (1991) call "integrated regulatory behaviour" in that the learning that took place was "personally important for a valued outcome" (p. 330). In other words, Ana was concerned with her own health and that of her unborn child. In the case of Mary, her concern was for the health and development of Laura, whereas Penny's concern related to her own bouts of ill-health. However, there was a trend in respect of their capability (and hence the time needed) to engage in self-directed behaviour to achieve the desired outcomes. Different perceptions of what outcomes were desirable had to be arrived at and acted upon. What Ana needed was the information to ensure she could remain healthy and produce a healthy baby. These conditions were already generally known and would have made the processes much easier. For Penny, there was a considerable degree of uncertainty over what was causing her ill-health and hence what she needed to do to retain good health. Mary's uncertainty was extreme, in that for some time she did not know much about Down Syndrome and its management and hence what she needed to do for Laura's well-being.

One factor common to all three stories was the need to master the technical language of medicine. All three women had some prior degree of health literacy, but still had severe problems in interpreting the meaning in practice of the specialist words they encountered.

There were four other factors that impinged on their learning, albeit to differing degrees. First was the ability to access their prior knowledge of the focus of their concern. For Ana, this involved recalling TV programmes and discussions with her friends. Knowledge about pregnancy and childbirth is widely available. For Penny, recalling the symptoms and causes of her husband's ill-health was involved, although these were not strictly parallel to her own. Mary evidently had no prior knowledge of or expectation of Down Syndrome, so understanding Laura's health needs must have been very demanding.

A second factor was access to the experience of other patients. All three were able to access the experience of other patients, albeit that it was of varying value. In the cases of Ana and Penny, their friends were of direct and valuable use. However, in the case of Mary, the experience of other parents of Down Syndrome children was divergent in quality and value. Where this advice was subject to evaluation, as was the case of that provided by the Down Syndrome Association, it was evidently of unequivocal value.

Third, the nature of access to the experience of medical practitioners affected their learning. All three received valuable help and advice from medical practitioners, in particular from nurses. The limiting factors on the availability of such help seems to have been the accuracy of the diagnosis of the health issue involved and the professional's knowledge of what should be done. Only in the case of Ana, where the issues of pregnancy are widely known, was there a relevant educational course available.

Finally, all of our authors needed access to written material. It was only in the case of Down Syndrome that detailed written material was available, although the language issues mentioned above probably made its use challenging. The widespread availability of Internet-mediated information in recent years will, it can be assumed, have overcome these issues.

The authors of all the three case stories, beginning from divergent starting points in respect of their self-educational skills, were able to engage in the learning that they sought, although in Penny's case, not yet achieving a satisfactory outcome. In later chapters we will return to these ideas on how a passage to full, early, and effective self-directed behaviour can be made easier.

References

Boodman, S. G. (2011, March 1). Helping patients understand their medical treatment. *Kaiser Health News*. Retrieved from http://khn.org.news/health-literacy-understanding-medical-treatment/

Bull, M. J. (2016). Clinical Report – Health supervision for children with Down's Syndrome. *Paediatrics*, 128(2), 393–406.

Cunningham, C., & Glenn, S. (2004). Self-awareness in young adults with Down Syndrome: I. Awareness of Down Syndrome and disability. *International Journal of Disability, Development and Education*, 51(4), 335–361.

Deci, E., Vallerand, R., Pelletier, L., & Ryan, R. (1991). Motivation and education: The self-determination perspective. *Educational Psychologist*, 26(3), 325–346.

Fong, J. (2010). Doctor–patient communication: A review. *The Ochsner Journal*, 10(1), 38–43.

Gordon, O. (2015, October 17). Living with Down's Syndrome: "He's not a list of characteristics. He's my son". *The Guardian*. Retrieved from https://www.theguardian.com/society/2015/oct/17/living-with-downs-syndrome-hes-not-list-characteristics

Graham, S. (2008). Do patients understand? *The Permanente Journal*, 12(3), 67–69.

Invisible Disabilities Association (IDA). (2016). What is an invisible disability? Retrieved from https://invisibledisabilities.org/what-is-an-invisible-disability/

Knapton, S. (2014, June 18). Nearly half of patients do not understand medical language. Retrieved from http://www.telegraph.co.uk/news/health/news/10909421/Nearly-half-of-patients-do-not-understand-medical-language.html

Lambert, H., & Rose, H. (1996). Disembodied knowledge? Making sense of medical science. In A. Irwin & B. Wynne (Eds.), *Misunderstood science? The public reconstruction of science and technology* (pp. 65–83). Cambridge, UK: Cambridge University Press.

Layton, D., Jenkins, E., Macgill, S., & Davey, A. (1993). *Inarticulate Science? Perspectives on the public understanding of science and some implications for science education*. Nafferton, Driffield, UK: Studies in Education Ltd.

Roth, W-M. (2014). Personal health-personalised science: A new driver for science education? *International Journal of Science Education*, 36(9), 1434–1456.

Stratford, B. (1994). Down's Syndrome is for life. *International Journal of Disability, Development and Education*, 41(1), 3–13.

4

PURSUING PERSONAL INTERESTS

Learning Through Hobbies

Many people pursue hobbies, particular interests or pastimes just for the pleasure of participation. At the same time, they are furthering their knowledge and expertise about their focus of interest. Hobbies may be defined simply as "regularly occurring interest-based recreational activities" (Zimmerman, 2015, p. 465), or more formally as "a systematic, enduring pursuit of a reasonably evolved and specialised free-time activity having no professional counterpart. Such leisure leads to acquisition of substantial skill, knowledge, and experience" (Stebbins, 2003, pp. 228–229). Either way, hobbies are leisure pursuits beyond the hobbyist's occupation that are interesting, enjoyable to do, and personally satisfying. Invariably, the hobbyist also gains knowledge and expertise (Liu & Falk, 2014) and in many hobbies, this new learning involves science and technology. The hobby of gardening, for example, is very much an enterprise based on science and technology. Although most hobby gardeners do not regard themselves as scientists or technologists, over time they accumulate a good deal of horticultural knowledge as they work out issues relating to planting, composting, pest control and so on, and accumulate a good deal of specific, if somewhat piecemeal, knowledge about science (Watts, 2015). In this chapter we present four stories about people who have pursued their own interests in different ways that have increased their knowledge and skills in particular areas of science and technology.

Pursuing a Lifelong Hobby

We begin with Richard and Michael, who turned their hobby pursuits into personal challenges to accomplish difficult tasks. Both men are the kind of hobbyist Stebbins (2007) would classify as "makers and tinkerers" (p. 8), and these

DOI: 10.4324/9780429434303-4

stories describe how each built his knowledge and expertise through satisfying accomplishments of quite different kinds.

Richard's Story: Building a Logie Baird Televisor

Richard began his collection of historical artefacts relating to light and sound technologies in his teens with the purchase of an Edison phonograph, and his interest in science and technology led him to become a physics teacher. His training gave him considerable knowledge in the relevant principles of physics associated with light and electricity. He also had a fascination for how things worked and the historical context of technological development of artefacts relating to sight and sound. Many years ago he founded a club for people interested in gramophones, wirelesses, and early television and he remains an active member.

> *A challenge I set myself in the 1980s was to build a working model of a 1920s Baird televisor (television set) using the same early technology that John Logie Baird had used. Baird televisors were the first television technology to be made readily available to the general public. Between 1929 and 1935, Baird television signals were regularly broadcast for short periods of time (usually 30 minutes each day) by BBC radio stations in the United Kingdom. Baird televisors could be used to view these signals. The picture was viewed as a 30 line image at 12.5 frames per second with a picture size about 4 cm by 3 cm. Although the frame rate was well below that required for flicker fusion (the rate above which flickering cannot be seen), Baird televisors were used by many in the United Kingdom. On February 28, 1928, using short wave radio sets, an historic trans-Atlantic transmission of television from London to New York was accomplished by the Baird Television Development Company.*
>
> *There were two aspects to the learning involved in this challenge. First was the understanding of the old technology; such as neon lamps, triode valves, and photoelectric cells. The second was the science understanding needed to make these old technologies work. Since there was no example of an original Logie Baird televisor available to look at, it was necessary to gather information about the design and operation of these machines from any published diagrams and photographs that I could find by searching through books on the history of television. I started by reading about how the Baird televisor operated.*
>
> *What is a Baird televisor?*
>
> *The original Baird televisor used a 50cm diameter Nipkov disc (a spinning disc for scanning images) with 30 small holes in a spiral around its outer edge. The television picture to be viewed was produced by a neon glow lamp or pilot lamp, whose brightness was controlled by varying the applied voltage. The neon lamp was placed just behind, and close to the edge of, the Nipkov disc. As the disc spun the light from the lamp passed through the holes on the outer edge of the disc. The moving spots of light produced a small rectangular area with 30 lines of illumination, viz. the picture. The*

modulation of the brightness of the neon lamp produced the light and dark areas of the picture. The picture was then viewed through a magnifying glass.

To build a Baird televisor, it was necessary for me to learn about the design and structure of a Nipkov disc. I made a disc of 51cm diameter from a sheet of mirror Perspex and carefully drilled the required 30 holes. Controlling the speed of rotation of the disc involved learning how to control the speed of the electric motor using a light dimmer switch. Original neon television lamps were no longer available; however, I found by experimenting that four neon pilot lamps connected in parallel could serve the same purpose. Research into neon glow lamps indicated that the range of voltages that could be applied to these lamps was between 90 V and 400 V.

The challenge involved in building a working model of a Baird televisor was two-fold. I needed to generate and record a video signal, and then devise a means to play it back and view the result.

Recording a video signal

Of course, today, 30 line television signals are not readily available. I was able to obtain an original photoelectric cell and use it to create a video signal. Research into photoelectric cells in the books available indicated that the minimum voltage required for these to operate was 90 V. To achieve this voltage, I connected twelve 9 volt batteries in series, thus producing 108 volts.

The photoelectric cell was placed directly behind the outer edge of the Nipkov disc I had made. A slide projector was used to focus a small image about 3cm wide (initially I tried a still picture) onto the front of the disc but opposite to the photoelectric cell. The 30 holes in the spinning disc thus 'scanned' this picture. The scanned signal passing through the holes was detected by the photoelectric cell on the other side of the disc.

The output from the photoelectric cell was then recorded on a tape recorder. However, I knew that the line-in (input) of the tape recorder had to be isolated from the high voltage applied to the photoelectric cell, or the recorder could be damaged. To achieve this isolation it was necessary to investigate the use of an isolation transformer, which was then placed in front of the line-in.

Playing back the video signal

The recorded signal could then be played back. It sounded like a repetitive noise. However, the signal had first to be amplified in order to provide the modulated high voltage that would make the neon glow lamps work.

Research using pre-solid state text books indicated that this amplification could be achieved using a triode valve. The line-out of the tape recorder was applied to the grid of the triode valve. The output of this valve then provided voltages ranging from 150 V to 350 V, according to the level of the tape-recorded video signal. This variation in voltage was applied to the neon glow lamp, and this caused identical variations in the brightness of the lamp. When the part of the Nipkov disc in front of the lamp was viewed through a large magnifying glass the scanned picture could be seen. Eventually a satisfactory 30-line 12.5 frames per second television picture was produced. The resulting model Baird televisor was able to be used in public demonstrations of this rare early technology.

In summary, to achieve my challenge, I had to learn about and understand a range of aspects of science and technology.

- *The structure of a Nipkov Disc and a means of controlling its speed.*
- *The voltage range of neon glow lamps.*
- *How to operating voltage and the use of a photoelectric cell.*
- *The structure and creation of a disc stroboscope.*
- *How to use of a triode valve to produce a modulated high voltage.*
- *Flicker fusion rate for human vision.*

The knowledge required to build the Baird televisor was new to me because I was initially unfamiliar with the technology. Fortunately, I was a physics teacher, so the challenge was not to learn new principles about science and technology, but rather to rediscover knowledge and understanding of the old technology. I knew about the focal length of a lens and how to connect lamps in series and parallel, but in most cases I had never learned about the technology of, for example, how to use photoelectric cells. This technology had long been surpassed by solid state electronics. Valve technology had also long been replaced by transistors and other solid state devices. However, the challenge in this project demanded the use of authentic original technology, and this meant I had to learn about the voltage range of the photoelectric cell and how to capture the modulated signal produced by the photoelectric cell when the televisor was in operation.

In some cases I made use of the advice of older technicians who had worked with these technologies in the past. These older, experienced persons could draw on their own knowledge, and could explain it to me personally. They had had first-hand experience with valves and photoelectric cells. However much of the required information was also readily available in books, particularly books relating to pre-transistor technology.

Today, my learning task would be much easier because we have access to the Internet. While much of the old technology is well documented, it is sad to see the passing of those persons who had first-hand experience in this area. However by constructing a working model of the Baird televisor I can now demonstrate to the younger generation something they would normally only read about.

Richard's challenge revolved around his desire to replicate, as closely as possible, the original technology developed by John Logie Baird, but many of the authentic components were no longer available and for those that were, how to use them was a challenge. Richard was delighted to succeed after a lengthy period of trial and error and improvisation. Because of his physics background, he was able to understand the relevant physics principles and improvise where necessary but his main source of information came from old books with descriptions and diagrams of the Baird televisor and how it operated. Being able to talk with old technicians who had actually used the components many decades previously was very helpful. These people have passed away but, as Richard pointed out, there is now an alternate source of information, the Internet. The website http://www.bairdtelevision.com/, developed by Baird's son and grandson, would have been of great interest to Richard. A current Internet search suggests, however, that most Internet information is about Baird and his achievements, rather than the details of how his televisor actually worked. To get those details, it is very likely recourse to those old textbooks would again be necessary.

Incidentally, it is worth noting the satisfaction that John Logie Baird himself felt on achieving his goal of transmitting images. He was an inveterate inventor and often the butt of jokes for his endeavours (Kamm & Baird, 2002). He is quoted as stating:

> I was definitely able to transmit the living image, and it was the first time it had been done. But how to convince the sceptical, hide-bound, select and exclusive scientific world? Would they admit that a wretched nonentity working with soap boxes in a garret had done something which many of them had stated was not possible?
>
> *(http://www.bairdtelevision.com/firstdemo.html)*

Michael's Story: Creating Complex Jewellery

Michael's background in science is linked to his training as a physics technician but, as he explains, his work place gave him access to additional avenues to pursue his great interest in making things with metal.

> *I first began making things as a young child. I can remember, at about the age of ten, being fascinated by trying to bend and twist wire into fanciful shapes. It was an exercise in imagination that perhaps was, in part, a refuge from being consigned to the back of the class, the fate of being dyslexic at a time when the condition caused me to be thought "dumb". I made these objects for the sheer fun of it, since they were not preserved or, indeed, much valued. I simply liked to play. The objects I made were small things – I recall, for example, making a tiny boat, with mast but no sails, to fit into a small light bulb from which I had carefully removed the end.*
>
> *At about fifteen, I was given apprenticeship as a physics technician at an Atomic Energy Research Establishment. I now had access to equipment and a greater range of materials that enabled experimentation in my spare time. I remember making a miniature cannon, playing around with soldering and joining pieces of metal together. I have worked as a physics technician all my working life, and have always been interested in exotic materials. Better machines allowed me to try to make simple objects by cutting aluminium or stainless steel – early on I found I could make shapes that I had never thought I would be able to do, by soldering and hammering and twisting.*
>
> *I first became interested in trying to make jewellery in 1973. The first thing I made was a stainless steel ring. What really started me thinking was that I saw a pendant*

made by a Brazilian artist in a colour magazine. I thought I would love to make something like that, so I bought some books on jewellery making. To my disappointment, the books were not much help. They were very limited in their usefulness as a practical guide – they told you how to do it in simple steps, but failed to point out the pitfalls. As a result, my early efforts were largely failures. I had picked silver as my working material as it was cheaper than gold, so mistakes were not too costly, but they were frustrating. I did not understand why soldering silver seemed difficult. I did not know enough about what one could and could not do with this metal. Silver-smithing and making jewellery is a very personal, hands-on art and I was not succeeding.

I knew I needed help. I found that the local technical college was running evening classes in silver-smithing. I went along to the open evening to see if it was what I wanted. I thought it was so I registered and for the next few years became a student.

I was very fortunate that the tutor in the course was a trained silversmith from Birmingham, the main jewellery centre in England. The first thing he asked me to make was a plain ring, as this is the simplest form of personal jewellery and a good starting point. I, on the other hand, wanted to make a pendant. Luckily, he was immensely supportive and helped me to understand the basic elements of construction. Silver is difficult to solder because it is a good conductor of heat, so when you apply the solder at a particular point, the point rapidly cools as the heat dissipates throughout the metal. The trick is to heat the entire object first, so that the solder can be applied at an appropriate temperature for an appropriate time. The books do not tell you that!

The pendant was to feature a ruby, for which we had a gemstone – but it was too large for the setting I designed. At the tutor's suggestion, we hit it with a hammer. My pendant was made with an irregular piece of ruby which exactly suited my wish to make individual, unusual designs, different from those you see in retail outlets. I was both encouraged and inspired.

With my tutor's help, I became more confident and more experimental. It was hard work, because the objects were made by hand, not on a mill. Some were beaten from flat sheets of silver – jugs, raising the sheet upward over a stake; bowls, beating it downwards into a well. I learned that there are actually three temperatures at which one can solder silver, and sometimes it is necessary to employ them all. Imagine, for example, that three narrow strips of silver are to be soldered to make a triangular piece for a brooch or a pendant. The first two will be joined at a high temperature, but if you try to solder on the third at that same temperature, the first solder will melt. So you need to use a lower temperature solder for the final piece. In similar fashion, I learned about cabochon gemstones: you need to file them carefully around the edge to make a seamless line with their silver setting. I was enjoying the work, and now began looking at magazines and jewellery shops, observing current styles and trends.

After a few years, the college abandoned the course. I and my fellow students, most of whom were, like me, in their 50s, knew that without the facilities of the college we would be unable to make many of the things we had practiced and perfected over our time there. We decided to form a club, so we could purchase the necessary tools as a

took up gardening in an allotment near her city, growing mainly vegetables for her own consumption.

> *While tending my allotment, I noticed a large black bee with a red tail. Having never seen a bee like this before, I looked it up on the Internet and discovered to my surprise it was one of over 260 species of wild bumblebees and solitary bees in the United Kingdom. The one I had seen was a Red-Tailed Bumblebee (Bombus lapidarius) which, like the domesticated honeybee, lives socially, with a queen, female worker bees, and male drones.*
>
> *I started to notice articles and TV commentaries on the concerns over the declining numbers of bees as pollinators. The Friends of the Earth, for example, state that bees are crucial to our economy – without them it would cost United Kingdom farmers well over a billion pounds a year to pollinate crops. However, honeybees were in difficulties and many wild bumblebees and solitary bees were also declining at an alarming rate. Three bumblebee species were already extinct in the UK. Currently forty-seven British bees are listed in the Red Data Book of threatened species. But none are protected by law* (http://www.foe.co.uk/sites/default/files/20-things-you-need-know-about-bees-15919.pdf).
>
> *While my interest was aroused, I only became more involved and focused when I noticed a request for volunteers to carry out a bumblebee survey for a Wildlife Trust and the local Co-Operative Farm. This had several advantages to learn more. It provided a manageable structured activity, some training, and an opportunity to share experiences with others. This would build on my experience as an educational researcher with a personal interest in general ecology, but with very limited knowledge of insect behaviour and identification.*
>
> *The Co-Operative Group Farm wanted to know if widening and/or seeding (with native flowers) the field margins would increase the number of bumblebees. During 2012 and 2013, volunteers were asked to walk a transect once a month and record bumblebee species. A day's training was provided initially, with a very limited amount of feedback, as we struggled with identification.*
>
> *A day's training each year proved to be far too little but it clarified the main aspects that identification involved. The task appeared manageable, as there are only 24 bumblebee species in the United Kingdom, but it was also important to distinguish between queens, female worker bees, and drones as well as cuckoo bumblebees. Cuckoo bumblebees invade the nest of other bumblebees and annoyingly look similar to their hosts. In practice, I found it was very difficult to identify the bees, which were constantly on the move. As a volunteer researcher, with little access to experts, there was the constant worry that incorrect identification would falsify the results.*
>
> *The first two years of my learning about bumblebees was characterised by piecemeal, ad hoc learning. As an independent learner, I did not have a clear strategy for learning because I did not know enough about the subject. There were occasional meetings with other volunteers and a Recorder's Conference organised by the County Council where*

the leader of our research gave a talk on the first year's findings. These activities provided snippets of complex information that did not always seem to fit together. I needed an overview on the subject.

The next turning point occurred when I found "A Sting in the Tale" by Dave Goulson (2013). This book provided a very readable humorous insight into bumblebees and related research. (Independent learning needs to be engaging and fun.) This book provided the structured knowledge into the life cycles and differences between British bees, which enabled me to see the relevance of new knowledge and to add new ideas logically.

At the same time, I came across the Bumblebee Conservation Trust, which was established by Dave Goulson in 2006 (http://bumblebeeconservation.org). *In 2012, the Trust started a national "Bees for Everyone" project to raise public awareness of the threats bumblebees face and to help rare bumblebees through active conservation work to safeguard, restore, and create habitats for them. Part of this was to ask for volunteers to walk a self-chosen transect to identify bees. This process was very close to that used by the Wildlife Trust in the Co-Operative farm research. I was delighted as I could continue to develop my new skills as well as choose an area that interested me. Consequently, I have been doing a "BeeWalk" in Aylestone Meadows since 2013.*

While I, personally, have been developing my knowledge and skills, the Bumblebee Conservation Trust has been adding to their provision for the general public and their volunteers. The trust provides monthly on-line newsletters about topical and seasonal issues. These are very good but do not provide the opportunity for interaction and feedback. I still wanted feedback regarding identification and for clarifying my knowledge.

Fortunately, the Trust has recently been developing these facilities through an on-line identification tool and conferences. The on-line identification tool can be used from a home computer. Individuals are encouraged to take digital photographs of any unusual species and upload them to their site. This "BeeWatch" tool provides relatively easy questions to help identify the bumblebee in the photo. Other Internet users are encouraged to check identifications and an expert will also look at examples and provide feedback, although the latter can take some weeks. Recently a new tool providing a self-testing set of photographs has been provided with immediate feedback.

To date I have attended two conferences, a year apart. Conferences have the advantage of practical workshops and opportunities to raise face-to-face questions but are relatively infrequent and not always accessible. However, the combination of on-line interaction with some opportunity to meet other people interested in the topic is ideal for me.

As I have become more competent in identifying Bumblebees, I have become curious about the relationship between different species and the flowers they pollinate. My new challenge is to identify the plants they use and discover why bee preferences vary.

I am now much more aware of my preferred learning methods. Therefore, I believe I am more able to tackle this new challenge more effectively. For example, I now know that going on a generalised course, in this case on flower identification, does not suit me. It would provide too much knowledge, much of which is not pertinent to what I want to know, and would be overwhelming. Consequently, I use a "need-to-know" strategy in which I only focus on flowers that bees visit and ignore other flowers. This is same technique that I used to learn differences between types of bees: photograph an unknown example that I want to know about and then research it though identification books, Internet, and friends.

Points of reflection

- *There is an initial ad hoc approach to learning because the learner doesn't know enough about the subject to learn in a logical fashion and does not know where to get help. The learner could give up at this stage.*
- *Total self-study needs a lot of self-discipline as other life demands can easily take over time. However, volunteering in a group provides structure, some feedback and motivation to continue.*
- *It is important to take full opportunity of increasing on-line support and chance to meet other interested learners.*
- *Expect to take a step-by-step learning process: I found a "need-to-know" strategy suits me.*

Tina's interest in bees was stimulated by observing an unusual bumblebee and wondering what it was. Gradually, and with the help of a humorous and insightful book (Goulson, 2013), her interest led to participation in a citizen science project to document the number and location of various kinds of bees. Projects described as citizen science are those in which "volunteers collect data for use in organised scientific research" (Bonney, Shirk, & Phillips, 2015, p. 152). There are many "data collection" projects involving volunteers, who often have no formal training (Bonney, Phillips, Ballard, & Enck, 2016). One issue often raised in such projects relates to the quality of data collected by non-scientists, and Tina, as a novice identifier of bees, soon became concerned about her own skills in correctly identifying the bees she observed. In order to become more effective (so as not to *"falsify the results"*), Tina sought additional training and information to ensure that her identifications were accurate. Now, after several years of thoughtful bee watching, her interest has expanded to the kinds of flowers bees pollinate, a natural extension of her participation in this pursuit.

Paulette's Story: Opaque Aquifers and Other Matters

In 2010, Paulette retired from her academic career in museum studies and moved to Mere, a small town in southern England. Her choice to retire to Mere was due, in part, to a love of the chalk lands of Salisbury Plain and the Jurassic coast of Dorset, that had grown over more than 40 years of driving from London to West Lulworth (which is south of Mere), especially for school holidays. Paulette has a science background in botany, but her move to Mere led her to an in-depth exploration of the underlying geology of the area, because Mere and its surrounds are shaped by a significant east–west fault line. Geology was a science new to Paulette, and this story is about her growing knowledge and understanding of its significance to the water quality and supply in Mere. Paulette's experiences led her into many different areas over several years and consequently hers is a long story. As background, Paulette wrote the following explanation of the geology (something she could not have done a few years ago) which will serve as an "advance organiser" for readers.

> *The geology of Mere*
>
> *The Mere Fault formed some 270–300 million years ago near the arid heart of the super continent Pangaea in the period between the late Permian and the earlier Carboniferous periods of geological time. At the time, the landmass of present day Africa was pushing northwards into Pangaea to join up with present day Europe and Northern America and there was much northern hemisphere mountain building. The basal Mere Fault was probably a lateral fault movement as there was none of the mountain building or subduction episodes associated with continental plates near it. The bit of land that Mere is on became covered with sedimentary rocks.*
>
> *Then, beginning 63 million years ago, the north Atlantic Ocean began to form near Greenland. As the Atlantic widened the old foundational Mere fault fracture was reactivated and the Jurassic and Cretaceous layers above it were disturbed as a dip slip movement along the fault from bottom to top took place over about a million years. Displacement and erosion gave us the landscape of today.*

Paulette's introduction to her story:

> *This is largely a meandering tale of the unfocused gathering over time of layers of understanding initiated by seeing amazing Jurassic and Cretaceous landscapes on the southern coast of England years ago. Aesthetic appreciation developed into a curiosity about how, over millennia, the places came to be as they were and how people, more recently, came to live in and around them. While I followed my career as a museum consultant and museum studies lecturer this interest was largely satisfied by walking over the land once or twice a year and having access to interpretative literature, maps, and television programs. The gathering was also guided by my early understandings of "how to be a research academic" and my subsequent use of tools such as libraries, statistics, visualization, history of sciences, and critical thinking.*
>
> *So, from the start, this was a fairly organised meander into a multi-disciplinary science area which encompassed geology, ecosystems, biogeography, anthropology, history, environmental concepts and straight out observational natural history; the sort of "cultural science" interest that is sometimes hard to interpret to others who might see the approach as trivial, superficial and "soft". In reality, I have found that selecting*

and following ideas, developments and insights from various disciplines over the years is a strength of the multidisciplinary approach to any many-layered situation.

Now, some years later, I am involved in a much more focused, pragmatic exploration of the somewhat opaque science of hydrology, the hard politico-science of water resource management and, also, the way that environmental NGOs [non-government organisations] function. This focus rests on the above interests.

The story begins

In 1980 I experienced the good fortune of being part of a week-long biological field trip, led by an excellent lecturer, to the nearby Purbeck coast. I learned a lot about investigating and describing from her. In 2010 I knew that, in Mere, I would be able to apply much of what I had learned from that experience as a hobby interest if I found it difficult to settle into this small unknown community.

Mere lies on a quite famous east-west geological fault of which I was ignorant. When I arrived I found that the town tended to interpret itself with regard to the fault with the main through street said to lie right on the fault so that the houses to the north were built on chalk and those living to the south had to struggle to make a garden on unforgiving clay. There are two streams (the Shreen and the Ashfield) that meet in the town. They rise on the high chalk land to the north and drain across the clay valley below to join the River Stour which eventually gets to the coast at Christchurch. Chalk holds water and clay does not. The streams were often explained as rising from a spring line which is not the case in Mere as the streams are not naturally ephemeral. There was no concrete sign of any fault line I could see and the line marking it on the geological map represents an area 50,000 times broader than the width of the paper representation. So I decided to try to find out where the Mere Fault actually went on the ground with the idea that I could, perhaps, walk along its route.

"Dead" (non-active) faults passing through towns had not crossed my mind. I had seen dramatic, picture-book quality fault patterns revealed by erosion on the cliffs of the nearby Jurassic coast. I had seen film of the Californian San Andreas fault which was described as due to the movement of two continental plates sliding past each other so that gate posts were no longer aligned with fences after earthquakes or tremors. Mere was not on the edge of a continental plate. It was an old settlement and would not have been built on uncertain foundations.

Throughout 2010 I was much involved in making a new garden from scratch, but now and then, well into 2011, I picked up information about the local geology by borrowing general introductory books to Mere and two books about Wiltshire geology and several general books about British Geology from Mere library. The Mere Historical Society books in the library were helpful as were the museum curators. I bought books that I thought were good and looked at the reference lists in them. I kept my eyes open when out and about and I tried to read the local landscape I walked over with growing fascination.

I found that if you want to get beyond general principles but not get bogged down in excessive detail, geology is a difficult subject to feel that you have some understanding of because of the time depth covered, the general slowness of the processes involved and the dynamic nature of changes over age long time. Keeping track of the time scale and what is thought to have happened is like constructing a movie in your head – personal visual thinking that you can "read" from and put into your own words flexibly is helpful. My local movie is still weak and often blank. To me, some general authors did not help me to visualise and seemed to be compiling a wilfully obscure, language-based account from other sources – unfamiliar words were not often explained and were opaque in the context given. Some more academic authors could tend to use the argot of a clique of those in the know. One thing continually puzzled me. The low lying clay land I lived on was said in all the books mentioning the Mere Fault to have been lifted 185 metres above the present day chalk I could see to the north before the lifted chalk surface was eroded down to the clay strata to be seen today. How would such a big physical event happen? Could you still see any signs of this event? It took three years for me to realise that it was not a singular event but a process that could come about centimetre by centimetre during parcels of thousands of years.

Additionally, geology is not an everyday topic of conversation. Thankfully, I found two friends, both academics as it happens, who did not mind geological conversations, usually as we were going for a country walk. Now I have found a source of such conversations in a University of the Third Age (U3A) geology group and discovered the Penguin Dictionary of Geology.

A stimulating event

In mid-summer 2011, the sources of the streams in Mere dried up. The picturesque walk beside the Shreen down Waterside, with bridges over the chalk stream to the houses, fishes in the water, ducklings and white flowered Ranunculus aquatilis, was just mud and stones. Similarly, the Ashfield, which rose near the satellite hamlet of Burton to the east, dried so that the mud of the sheep wash, restored by the Mere Historical Society, and the millpond showed the footprints of the birds who usually swam there. There was little water in either stream bed up to the point where springs were trickling near their confluence.

There was a huge, protesting Parish Meeting. Wessex Water (a Malaysian company) abstracted (pumped) consumption-supply water from Burton Field below the contour where both streams arose and midway between them. Their representatives came, as did representatives of the Environment Agency, to whom water companies are in part responsible, and County Wildlife Trusts who had an environmental interest. It was explained that it was a drought year, not much to do with abstraction, but the residents were not pleased. Abstraction would continue as the company was legally obliged to provide water to households. It was explained that the abstracted water was sent from a reservoir up on the chalk down to Mere, to a large town to the south and several nearby villages. A very informal Mere Rivers Group began to coalesce around residents who lived beside the top of the Shreen on Waterside. They would announce a meeting and people, like me, would just turn up. The local paper published accounts and the National Press eventually reported the news.

I decided to find out how the Mere Fault actually worked and to use water to do so. Towards the end of 2011, I asked the Mere Museum voluntary curator if she would let me put a survey about wells, springs, ponds and seepage points in the museum and she agreed. Meanwhile I compiled historical accounts of the same. I was now stepping into the field of hydrology.

The survey response was very high and replies were often very detailed. They included accounts of 30 wells, 23 springs, 12 seasonal seepage points, 5 spring and seepage points deliberately fed into drains and 27 ponds. By Easter 2012 I had put the data on to a 1:1,200 map of the parish, provided from museum stores, on which individual houses were marked. We called it "The Water Map of Mere".

Motivations for my continuing interest in geology and hydrology

Mains water came to Mere in 1914 and was mainly sent from the Burton Field pumping station to four surrounding farms. Many houses were not connected to the water main, or the sewage main, till after World War Two. People not much older than me showed me dipping stones on the stream banks where they used to stand to collect water in a bucket for their mothers. A lady took me to a set of steps in a hole in her garden and revealed a dipping point to an underground spring on its way to the main stream. It was written into her title deeds that she should always allow her neighbours access to the dipping point. Some people with wells inside their houses had glassed them over as a decorative feature. It became very evident that the idea of the streams and "our water" was an important textural part of the sense of place that Mere people felt. People told me their stories and it seemed important to me to acknowledge their ideas and feelings. Plus, they were interested in what I was doing and that was encouraging.

Meanwhile, the survey data was indicating that the wells were mainly to the west of the town on deep chalk and that the water levels in them were historically around twenty-five metres, so giving an indication of the usual historical height of the water table. The springs were mainly in a distinct area to the east near the river courses where people could dip for water instead of digging a well. I began to visualise the fault line as very clean to the west and in a very jumbled up "earthquake area" to the east. The geological map revealed that there were two minor faults originating in the east on the main fault. They branched out to the sources of both streams from the jumbled up area. My "fault movie" was forming. Water and the fault were now inextricably linked in my mind.

In August 2012 Mere Museum put on an exhibition about water in Mere with the Water Map of Mere as an exhibit. There were also historical documents about sinking wells; the water supply and early twentieth century council infrastructure; geology; the local mills and industries which had been dependent on local streams and water supply. The Museum followed that with a small display about the 2012 situation in the clock tower in the centre of town. The general population was becoming familiar with the language with which to talk about this environmental problem.

Meetings with the water company and the Rivers Group continued as they do today. They were held in the recreation ground. After the mention in the national press the company appointed a consultant as an intermediary. She bombarded us with emails for around eighteen months till she left. Meanwhile, we got to know the environmental scientists who monitored the streams and they showed us how they did it. From this time a member of the Consumer Council for Water came to our meetings and we came to learn about OFWAT, the national water company regulator. OFWAT approves the business plans of all water companies every five years. The present period lasts till 2020. We tried to work out how the company functioned with regard to Mere. We researched how the abstraction level was reached. We sampled the rivers for organisms for the Wildlife Trust Riverfly Partnership program. So there was a lot of learning on all fronts and luckily the core group could share the job out and explain bits to each other.

There is now a core group of around half a dozen people, still sharing learning and activities. Mere rivers are protected under EU Habitat Directives. OFWAT must respect those directives. Maintaining the biodiversity which depends on their flow will be their saviour. It may mean that the abstraction licence will be reduced. Ninety eight per cent of all chalk streams in Europe are in southern England so, technically, they should be safe.

Sometimes the water company says that both streams are winterbournes – streams that flow from an overflowing aquifer in the wet winter months and dry up in the summer. We (especially those older people born in Mere) say that they are permanent streams, excessive abstraction has lowered the seasonal flows and that biology and history support us. However, the Rivers Group now hears voices from both the Environment Agency, who monitor the water company, and Wessex Water suggesting that the flow of the greater stream comes from an underlying greensand strata – not the chalk. If so, this would mean that the river was not well protected and pumping water out of the chalk aquifer cheaply, instead of collecting surface water in reservoirs, could continue. Without drilling down to examine a core – who knows? It could be our word against professionals. We have core data from road works nearby that indicates that the source is on the lower levels of the chalk aquifer. Along with the Wildlife Trust, we asked a Wessex Water environmental scientist for a ground water reference which we could look up. He gave a 1985 reference with a lot of American material in it and their record on aquifers does not seem to be great.

The evolved situation

In 2016, the streams partially dried in summer. We were given a grant to help with clearance of branches overhanging the Ashfield, an article about the abstraction problem in the South West Wiltshire newsletter of the national Campaign for the Preservation of Rural England (CPRE) was brought to the attention of the Town Council by the Parish Clerk, and there were staff changes at Wessex Water.

By early 2017, the core Mere Rivers Group realised that they would need a more formal identity in order to handle money and to decide how to deal with those who now wanted to join the group out of a general or specific interest. In spring we had a public AGM in the biggest hall in town with tables scattered round the floor, wine for the audience and six core members popping up to deliver a 10-minute update on various topics. The season was dry; the streams had begun to diminish in depth and flow; and silt from the winter by-pass run off was covering the ranunculus which had no lovely white flowers. Additionally, there were plans to build 1,500 houses in Gillingham, the town to the south (water to be supplied from Mere Source abstraction) and developers were also besieging the Town Council. We had a high turnout.

I followed the meeting with a geology walk around the formations responsible for the permanent streams which was attended by people from the AGM, the regular Mere Walking Group and members of the Mere Historical Society (the walk was also commemorating a local geologist who took the same route for a Geological Association field day in 1937). Suddenly lots of people in Mere were talking about the water supply, "our" streams and the geology – their "sense of place" was becoming more complex.

At a recent core group meeting we decided to stay small but with active workshop affiliates. These new groups illustrate the varied motivations people hold with regard to ecological or environmental interests. Wildlife sightings sent in by email will help build a database showing biodiversity; river bank management to allow light onto the Ashfield to add to the amenity value of the stream and encourage the insects which fish feed on; the difficult co-ordination of riparian owner activities on the picturesque stretch of the Shreen where the gardens of houses come down to the bank, may be tackled; a campaign of water saving will happen and the core group will continue to reason for a reduced abstraction licence on the grounds of the rarity of chalk streams and the biodiversity supported by Mere's streams.

With politeness and persistence, we have come a long way since 2011. At the recent meeting the leader showed an email from Wessex Water's Head of Water Resources showing the reduction in export from Mere when the flow was low or the ground water reached a trigger level and the rivers still had some water in them. The marvellous sight of water flowing out of the chalk rock is still there.

Who facilitated this learning journey?

- *those who wrote environmental TV programs and general books on the environment;*
- *those who funded fees and grants for mature students;*
- *a person who taught me practical ecological field work;*
- *general authors on geology;*
- *conversational friends;*
- *the local Library;*
- *members of Mere Historical Society, past and present;*
- *the Curator and Curator Emeritus of Mere Museum;*
- *the scientific staff of Wessex Water;*
- *people of Mere; and*
- *the Mere Rivers Group.*

Paulette's story presents a strong learning curve extending over several years built from a range of resources, including printed material like maps and books, people with relevant expertise and who supported her efforts to learn more. She quickly recognised the multi-disciplinary nature of the water supply issue in Mere and the powerful social underpinnings to the problems that arose from the diminished stream water flow in 2011 and again in 2016. Paulette's endeavour to understand the water flow issue and its relation to the underlying geology led her to use her research skills to collect data from long-term residents and carefully map the locations of wells and springs. This had two significant outcomes. On the one hand, it assisted Paulette to create a "movie" in her head to understand the geology, and on the other hand, it provided reliable historical data that established the natural permanency of the streams and usual water level in the aquifer. Despite these sources of information, the company supplying the water to the community argued for drought seasons and intermittent stream-flow to explain the dry streambeds, and continued their level of abstraction to supply and sell water not only to Mere, but to other towns and villages as well. The display in the town's clock tower of Paulette's map and historical data she had collected, including photographs and personal stories, also served as a means of engaging and informing other Mere residents about their water supply. Paulette and a few other established residents have become a well-informed group about Mere and its water and continue to advocate for reduction in the current abstraction license in order to maintain the health of the unique chalk streams. At the end of summer in 2016, when the stream beds were again dry, this group was able to arrange representatives of the Environmental Agency and others to inspect the streams, hopefully as a precursor to recommendations for future abstraction levels.

Commentary on the Four Case Stories

The four stories presented here are quite diverse. Richard and Michael described lifelong hobbies involved with personal challenge to master their topic, Richard in terms of preserving and restoring old light and sound technology for future generations to appreciate and Michael in constructing creative and increasingly complex art pieces. Neither was interested in monetary rewards for his personal achievements and clearly their hobbies are outside of their working life, although their chosen careers provided some background knowledge and equipment to further their hobby activities. In contrast, Tina and Paulette focused on matters stimulated by events external to their personal life, contributing data to issues of concern much wider than themselves, to a scientific study of bees and the maintenance of healthy chalk streams, respectively. Both pursued these environmental activities in their own time for their own interest, so both can be described as hobbies according to definitions noted at the outset of this chapter.

Liu and Falk (2014) listed four key characteristics of the nature of hobbyist learning: "(1) the situated, real-world nature of what is learned; (2) the social, community aspect of the enterprise; (3) the development of expertise or mastery; and (4) the important role played by interest and intrinsic motivations" (pp. 345–346).

All of these characteristics are relevant, in somewhat different ways, to the case stories presented in this chapter. In terms of the first of these characteristics, for example, Richard built a working model of a real-world, albeit historical, artefact, and Michael made usable silver pieces. Tina's and Paulette's stories are clearly real-world issues. The second characteristic is also important, but for different reasons. Richard and Michael both sought assistance from others with the specific skills they needed to accomplish their own, personal challenge, and each has been instrumental in the formation of clubs relating to their hobby. Tina and Paulette joined local groups, not simply to achieve something personal but to contribute actively to the environmental issues at hand. When she began, Tina was concerned that she had insufficient skill in identifying bumblebees and tried some courses and websites to improve her skills and hence contribute data more confidently. Paulette felt her way gently to begin with, consulting people from the local museum for local information, and gradually expanding her knowledge with help from those with more ecological, geological, and hydrological expertise. However, she then contributed to the community's knowledge about Mere water by leading the clock tower presentation of her findings, and she continues to participate in a small Rivers Group and leads geological walks to help others appreciate the local landscape.

It is notable that both Tina and Paulette began their learning journeys through curiosity. Wondering what kind of bee Tina saw in her allotment led her first to "look it up" on the Internet, then to actively study bees and now, having learned a considerable amount about them, her interest has turned to the flowers bees prefer to visit and the reasons for those preferences. Paulette's attention was drawn initially to the low water levels in Mere streams, and pursuing her curiosity about that has led her to involvement in, as she relates, geology, ecosystems, biography, anthropology, history, environmental concepts, and straight out observational natural history. Such journeys into new but related fields are quite typical for learners who begin a search for science information through curiosity. Morris (2015) describes

> how readily simple curiosity develops into a quest for greater understanding … how discussions based on real-life questions draw simultaneously on ideas across the various scientific disciplines … and lead beyond the natural sciences into ideas in religion, sociology, history, economics and all the other humanities.
>
> *(p. 8)*

For over a decade Morris (2015) has held informal discussions with groups of "ordinary citizens" with little science background who wanted to learn "something about science", and he gives numerous examples of how the trails of discussion can begin in one place and end in quite another. Morris' discussants simply wanted their curiosity satisfied, and they then moved on to different questions, unlike our storytellers, who have not only pursued their topic in depth, but have become sufficiently expert in it to help others to understand it.

All of the case story participants used a range of resources to develop their knowledge and skills. Books were important to all, as were significant others. The Internet was of most use to Tina because as more supportive resources were developed she was able to get rapid feedback on her bee identification and other concerns. The Internet was less relevant to Richard and Michael because their lifelong hobbies began before computers were common and there was no Internet. Both are now masters of the relevant technical aspects of their hobbies, but there was a great deal of trial and error in trying to make things work. This is a common feature of many hobbies: The determination of "what works" was also prominent in the stories told by gardeners (Watts, 2015), who learned from doing and consulting others with more experience. Tina and Paulette learned by doing, too, although there was some reliance on others to give them confidence in the accuracy of their identifications/interpretations. Tina's participation as data collector in the citizen science project provided a structure that assisted her to deepen her knowledge of bees through attending courses, meeting others, and communicating via the Internet. In contrast, Paulette did not volunteer to participate in a particular project, but she also could be described as a citizen scientist in a "community science" project, projects that are developed by members of the public about a local health or conservation issue (Bonney et al., 2016). Paulette's own interest and desire to understand the Mere Fault and its relationship with the water supply led her to collect data using a scientific approach, and the data she collected became a useful resource for others, including scientists. She became an activist for the conservation of Mere's streams.

Finally, each of the stories reflects the role played by interest and motivation to succeed in the participant's chosen challenge. Participation in hobbies results in what Stebbins has called durable benefits, outcomes that include one or more of "self-actualization, self-expression, self-enrichment, re-creation or renewal of self, feelings of accomplishment, enhancement of self-image, and enduring tangible products of the activity (e.g., a painting or cabinet)" (Stebbins, 1980, p. 413). Corin, Jones, Andre, Childers, and Stevens (2017) described in some detail how hobbyists (birders and astronomers) made extensive use of science related organisations as they built their knowledge and experience about their hobby. Reading each story reveals the participant's determination to move forward in knowledge and expertise and exudes the sense of "durable benefits" that come with competence in the field. Other case stories reflect similar self-direction and sense of achievement, a theme that will be returned to in later chapters.

References

Bonney, R., Phillips, T. B., Ballard, H. L., & Enck, J. W. (2016). Can citizen science enhance public understanding of science? *Public Understanding of Science*, 25(1), 2–16.

Bonney, R., Shirk, J., & Phillips, T. B. (2015). Citizen science. In R. Gunstone (Ed.), *Encyclopedia of science education* (pp. 152–154). Dordrecht, The Netherlands: SpringerReference.

Corin, E. N., Jones, M. G., Andre, T., Childers, G. M., & Stevens, V. (2017). Science hobbyists: Active users of the science-learning ecosystem. *International Journal of Science Education, Part B*, 7(2), 161–180.

Goulson, D. (2013) *A sting in the tale*. London: Jonathan Cape.

Kamm, A., & Baird, M. (2002). *John Logie Baird: A life*. Edinburgh, Scotland: National Museums of Scotland Publication.

Liu, C-C., & Falk, J. H. (2014). Serious fun: Viewing hobbyist activities through a learning lens. *International Journal of Science Education, Part B*, 4(4), 343–355.

Morris, A. (2015). *Getting to grips with science: A fresh approach for the curious*. London: Imperial College Press.

Stebbins, R. A. (1980). "Amateur" and "hobbyist" as concepts for the study of leisure problems. *Social Problems*, 27(4), 413–417.

Stebbins, R. A. (2003). Hobbies. In J. Jenkins and J. Pigram (Eds.), *Encyclopedia of leisure and outdoor recreation* (pp. 228–229). London: Routledge.

Stebbins, R. A. (2007). *Serious leisure: A perspective for our time*. New Brunswick, NJ: Transaction Publishers.

Watts, M. (2015). Public understanding of plant biology: Voices from the bottom of the garden. *International Journal of Science Education, Part B*, 5(4), 339–356.

Zimmerman, H. T. (2015). Hobbies. In R. Gunstone (Ed.), *Encyclopedia of science education* (pp. 465–467). Dordrecht, The Netherlands: SpringerReference.

5

LEARNING TO HELP OTHERS

The desire to assist others in their learning is a powerful one for many people, especially parents of school-age children. The "other" could be a child who has a need to know outside of school, or a person visiting a scientific institution such as a zoo or a science centre. Those who travel on environmentally-based tours are people who are interested in the science of a particular place, usually facilitated by a "tour guide". Those who "explain" in scientific institutions are variously called "explainers", "educators", or "docents". Those who explain to children are often called "parents" or "grandparents". In this chapter we present five case stories of adults whose wish was to help others to learn.

Helping Children

The importance of parental help for their child who has an assignment or homework about science is generally well recognised. Several authors have found that what happens in the home is critical. Parents influence both student attitudes and motivation; the influence of their mother in this regard is significant (Bandura, Barbarinelli, Caprara, & Pastorelli, 1996; Gottfried, Fleming, & Gottfried, 1998; Juang & Silbereisen, 2002). Given this generally positive trend, therefore, how well equipped are parents to assist? Parents have a reason to learn and are highly motivated, but how ready might the parent be in terms of prior experience and knowledge?

The science that is taught in schools today makes greater demands on learners than when their parents were students, particularly in high school, but increasingly at the primary level also. Early experience of the Internet and curricular expectations of inquiry learning result in project-based tasks that require some level of research. The complexities of school science are, however, greater than

DOI: 10.4324/9780429434303-5

those of 30 years ago and parents may be unfamiliar with the content and the context. Many, therefore, feel that their own knowledge is inadequate and seek further knowledge to help their child. The two case studies that follow exemplify this dilemma and describe how the two parents, in very different ways, set out to resolve it.

Paul's Story: Pop-Up Dinosaurs

Paul is the father of two daughters and takes a strong interest in their academic achievement. He himself has degrees in archaeology and museum studies which, as will be seen, helped to direct a possible pathway for his daughter's learning. His understanding of how to find information and of the value of a broad approach to exploration of a topic stood him in good measure during the experience he describes below.

> *My family comprises myself, my wife and our two daughters (15 and 8 years respectively when the activity recorded here took place). Our two daughters have spent their education years at a public elementary school and high school.*
>
> *Over the years the children have been given a variety of homework or project-based tasks by their teachers at school. At elementary school these usually take the form of homework sheets whose principal focus is on Literacy and Numeracy. However, the projects themselves can encompass a variety of subjects which will include science or technology-based tasks. Our youngest daughter, who we shall refer to as Emma, was given a third grade project by her school, which was to devise a pop-up book on the subject of dinosaurs.*
>
> *Emma's Dinosaur Project was partially carried out during school hours and partially ran alongside her regular homework. The more she could find out about the subject outside of school, the more she could contribute to school lessons. She came to us, her parents, for assistance. Perhaps the first thing I noticed when I was shown what the third grade students already knew about this growing subject was how much knowledge had been accrued – even at a basic level – since I was a child. A handful of creatures from my childhood were now engulfed by an army of dinosaurs with exotic sounding names. It put us "on the back foot" in that it seemed we were unlikely to enhance our daughter's work without having first to do some of our own research.*
>
> *The choice of where to find the source material was not as easy as I thought. A few of the BBC Television series "Walking with Dinosaurs" DVDs were on the shelf in the living room. They were devoured quickly. A scrambled search for a good old-fashioned book revealed an alarming lack of material other than unhelpful fictional work.*
>
> *Of course, the Internet did not disappoint. I found new several "new" species of dinosaur, discussed with Emma various theories about life cycles and habitat and was able to come up to speed reasonably quickly, provided of course that I took websites such as Wikipedia as a starting point and not as the last word. But it wasn't quite*

> *enough for Emma so I took her to a Natural History Museum. My parental response to Emma's call for assistance seemed to me to be a straightforward one. However, the visit turned into a wider educational experience. Before a dinosaur was even approached, the tactile seduction of the museum interactive displays in a different section of the building had captured Emma, eight years old at the time. I could not stop her from running to the geological area and engaging with the displays.*
>
> *It was some considerable time before we found our dinosaurs. Of course, the visit did wonders for our given project. Emma's pop-up book was delivered, overflowing with her interpretation of all she had seen and read. I took her again and again to the same museum at her request. She has yet to use the knowledge gained from the geological exhibits that diverted her interest in dinosaurs at the time of our visit in any homework she has been set.*
>
> *I wonder, if it had not been for this museum visit, I might have felt intimidated by the vastness of the subject matter and would have relied solely on the Internet for the quickest of kills.*

Paul recognised quite quickly that his own knowledge of dinosaurs was insufficient to help Emma and consulted conventional resources available at home for a "quick fix" of the problem. It is striking, however, that Emma was not satisfied with the information that Paul found in books, film, and on the Internet. Paul evidently felt that this search had yielded valuable – if somewhat minimal – information, but Emma wanted to know more. Despite the books at home and the information on the Internet, what was available from these resources was unsatisfactory. At this point, Paul called on his previous experience to provide for a visit to a Natural History Museum. This resource differed from their previous research in that it offered a tactile, interactive, concrete experience. Paul also allowed Emma to explore freely at the museum while he himself added to his own knowledge about dinosaurs that he had gained from those earlier resources. Repeated visits to the museum added to their joint knowledge and further increased Emma's interest in natural history.

Thus Paul brought to bear his understanding of where information might be found and what kind of information was needed, and he acted on that knowledge. They had evidently been to the museum before, but not to the "dinosaurs" section. Nevertheless, Paul intuitively thought that the visit and the concrete experience would help both him and Emma to understand the topic more fully. Despite the attraction for Emma of other exhibits at the museum, in particular those concerning geology, the eventual analysis of the dinosaur section enabled construction of material suitable to contribute to a class discussion built around a pop-up display to take place at school.

This project was evidently somewhat open-ended, requiring Paul to help with selection of information and its synthesis. The learning was really enhanced because of Paul's particular educational background and his willingness to accompany Emma to this most appropriate museum.

Liz's Story: Science for Mothers

Liz is also the parent of two daughters, and had been used to helping her children with their homework when they were in lower elementary school. She had found, however, that her ability to assist with science work was becoming more problematic as her older daughter progressed through school. Her own science background was limited to middle high school. A feeling of unease and apprehension had grown, to the point where she felt she would need help. A fortuitous advertisement for a short science course for mothers of high school students caught her eye and she enrolled in the informal program, which was of six weeks duration, one morning a week (for a full description, see Stocklmayer, Durant, & Cerini, 2011). The program took place at the local science centre. This is Liz's story.

> *I am the mother of two girls. At the time of the program, my elder daughter was in sixth grade and my younger in second grade. Although it was aimed at the mothers of high school students, I knew that I would be needing help very soon. In our family I am the parent who helps with the English and Humanities homework, while my husband is the parent who helps with the Math, Science, and Technologies homework. Basically we play to our strengths. Unfortunately Dad isn't always around when homework help is needed.*
>
> *I'd had a love of science in my first year of high school. It was most likely because my teacher was brilliant – Mr Smith. I still have copies of some of the work we did with him. He left the school very unexpectedly, and I struggled with science after that. It may have been because I was totally disengaged, or it may have been because I didn't understand the work. I did keep my love of learning about science through television shows like The Curiosity Show, Towards 2000, Quantum, and Catalyst. I loved taking my children to the local science centre.*
>
> *When the opportunity arose for me to learn how to help my children with their science homework through the program, it was too good to refuse. It was a class for mothers, run by three female scientists. It felt like it would be a safe learning environment. The classes were engaging with fun activities, clear notes (that I loaned to someone and didn't get back – hopefully the person is using them), special guests who did experiments and demonstrations with us, and the opportunity to ask questions without being belittled by classmates. Extracting DNA from strawberries was fascinating, and who doesn't want to make a strand of DNA with sweets (candies) and liquorice? These activities were designed so that we could replicate them at home, so the practicality of the course was invaluable. The cloud in a bottle was fascinating, as was the candles and carbon dioxide experiment.*
>
> *We had homework to do each week. Some weeks were more challenging than others, but there was a real sense of achievement when I was able to do it. The week we had a chemistry worksheet I even managed to combine elements correctly and*

balance equations. I had never been able to do that in high school. It gave me the confidence to know that if I could do it myself, then I'd be able to help my daughters.

When we reached the end of the program, there was a sense of both elation and loss. I had loved taking part and learning so much, and was happy that I felt confident to help my daughters, but wanted to keep learning more. I also couldn't wait for an opportunity to put my newfound knowledge to the test. When it came, it was a tough one, a story about a superhero in the form of an element chasing down a 'baddie' in the form of an acid, and what would happen when the two collided. Of course my elder daughter chose mercury and sulfuric acid. We did the work, checked and rechecked my notes, researched on the Internet, but something wasn't quite right. Thankfully the last thing our course facilitators had said was to contact them if we needed help. How often do you have scientists you can call on for extra help?

Ions were the subject of part of a recent ninth grade science assignment that my younger daughter was doing. I was able to help with most of it, even though it had been many years since the program ended. I found the mostly blank Periodic Table that I had laminated, and the basic element cards that showed each element's valency, as well as a whiteboard marker, and away we went, attacking each and every question with vigour. When we arrived at the final question, I was not quite so good at conveying what was meant to be happening. Thankfully a facilitator once again came to the rescue. With a quick visit to her office and an experiment or two later, we were able to deal with the rest of the assignment. I loved being back in a learning environment even more than my daughter did.

I was thrilled when the same daughter said they were looking at DNA. I really wanted to extract DNA from strawberries with her, but she was doing that at school the next day. Instead we visited the confectionery aisle of the local supermarket so we could build that edible DNA strand. It was built, photographed, and then devoured.

Over the years I've been spurred on to keep learning and trying new things. I wanted to keep on being able to balance chemical equations, so I'd do worksheets that I found on the Internet. After I inadvertently gave my notes away, I bought a couple of books to help me. I found the University of Nottingham's Periodic Tables of Videos with Prof. Sir Martyn Poliakoff and Brady Haran. That led me to Sixty Symbols. I still don't quite "get" physics, but I'm trying.

I stumbled across "Futurelearn" a few years ago, looking for information to help with yet another piece of homework. They had a free course called "Kitchen Chemistry". It had been put together by the University of East Anglia. It was almost a continuation of the chemistry work we had done in the program. Note to self: find a physics one. I've also taken part in a couple of citizen science projects. One I loved was classifying galaxies with the Galaxy Explorer project. It was interesting looking at the different types of galaxies, and thinking about how far away they are, and how amazing our universe is.

I have enjoyed learning more science since the program ended. Chemistry has appealed more to me since I felt such a sense of achievement in finally understanding the basics and I wanted to really consolidate what I'd learnt over those few weeks. I had a basic understanding of biology at high school so I was relatively comfortable helping my daughters with that. Pulling apart an onion and looking at the cells under the microscope I received for my 13th birthday was always fun. Physics – well, I'll keep looking for short courses and information to help me learn what I need to know.

Liz is a concerned parent who, like Paul, had been accustomed to being a resource for her children in many areas of their learning. As they grew older, however, her own lack of knowledge about science led to an increasing worry that she would not be able to assist for very much longer. The program was opportunistic – she might have chosen to enrol in an online course (as she later did) or in a more formal institution. The important point about the program, however, was that it set the necessary knowledge in a helpful framework and was pitched at the appropriate level. Once equipped with the knowledge and skills she wanted and needed, she could continue to be an important resource for her daughters in the area of science learning.

Liz believed she needed to know about a range of science topics over the long term, at a level appropriate to high school science. She had high commitment to the learning process but did not feel confident in exploring science knowledge and applying it to the task at hand. Liz was aware that difficulties lay ahead, as her ability to interpret the knowledge she sought diminished as her daughters grew older. Recognising that her self-belief was waning, she decided to be proactive in remedying this. Once armed with the basic information from the course, she continued to seek new knowledge and it became a very personal journey, culminating in involvement in a citizen science project. The presence and availability of mentors was important to her and continued over several years. Her learning in science has persisted: it is curiosity driven and is conducted within the areas of science with which she feels comfortable. At this point, therefore, it has become almost entirely personally motivated.

In contrast to Liz, Paul's belief in his own ability to help his daughter was high. He needed to know about the dinosaur topic, in a very immediate sense, and his prior experience was a critical factor in the success of the venture. Importantly, he had both commitment to the learning process and confidence in setting about finding a solution. He was able to help Emma with her project by considering where the resources they consulted had failed to meet her expectations. His own learning was short-term but fulfilled her need for help at that time.

The Explainers

In many areas of informal science outreach, explainers, guides, and helpers are employed to talk to the public. Taking the job of an explainer or a guide is often motivated by an early interest in the kind of science to be explained. These people often have limited science backgrounds, but have an interest in science and in being part of a wider community. The next stories in this chapter reflect different ways of helping, and illustrate different ways in which the narrators sought to increase their own background knowledge. Although the three stories described here also differ in the kind of place where the explaining occurred and in the outcomes for the people concerned, the motivation to share knowledge with others was common to them all.

Tiki's Story: Interpreting Plants

Tiki pursued a degree in computer science and physics, but she had had an abiding interest in plants from when she was a small child. It was a desire to "explain" that led to her becoming a science communicator and, eventually, an interpreter at a major botanic garden.

> *I've always loved plants, and knowing stuff about plants. My pre-teen years were spent living on an orchard farm that still had large areas of natural bush. My dad was fascinated by anything he could grow that was edible. My mother was a big believer in self-sufficiency and in making systems work their best. My botanist grandfather lived there too, when he and Grandma weren't travelling to interesting nature reserves around the world. So I grew up with a very random set of knowledge about plants gained through random conversations while tagging along behind one adult or another, at least half of which would have been mostly accurate. The natural world was my world – I tasted anything that seemed like a good idea to chew on (and a few things that weren't), I ran, slid, and rolled on all kinds of pasture, I hid in leaves and built little houses out of branches, I breathed deep of the forest and learnt to recognise the different forest types and seasons by their scents.*
>
> *My university education ended up being in science – but in physics and computer science. I'd been well steeped in the idea that you serve your community as well as yourself, and spent my university years volunteering at the university radio station. In radio I began at the afternoon talk show, where I absorbed a lot of random information about a wide range of things in the world, whether I cared for them or not. I had to learn the skill of becoming an instant mini-expert in something – not for myself, but because I had to write a background sheet and questions for the show's host. The host would perform several interviews in an afternoon, often with well-qualified guests and the host had to know how to steer the topic towards what was interesting, topical, and understandable for anyone listening, who probably didn't know anything about the subject matter. With that interview pace, they needed the research done for them. This was before the Web had really come into its own, and Google was still some years from being invented. So, we had other means. And mostly, those were people such as the radio background team and publicists. And interviewees themselves were usually quite happy to discuss the topic beforehand so that everyone was "on the same page". You just had to master the art of asking them good questions, and the art of listening well to the answers.*
>
> *When I left university and the radio station in 2000, I saw an opportunity to volunteer at the Botanic Gardens, as one of their guides. I was looking for a next place*

to volunteer, and I'd never stopped loving plants. This turned out to be a lot more intense than I'd been expecting. The Guides are the Garden's front end – they operate the Information Centre, they lead free public walks, they are many visitors' only contact with the Garden's Authority. As a consequence the training is very thorough. Visitors come to a botanic garden expecting a minimum level of accuracy from the people they talk to, and they don't distinguish easily between a degree-qualified employee and the volunteers just helping out in their own time. Suddenly, I had to formalise and deepen a lot of the knowledge I'd vaguely counted myself as having. Instead of following people and asking questions, other people would be following me.

The first part of this was the training provided. Our training course provided us with a lot of reference material that explained the basics of botany, and of ecology as it related to the Park environment. There was a fair emphasis on interesting facts, because those would be of regular use to us, but every fact had to be set in its context at least for our understanding so that we could use it correctly. I still have the training materials from that course, because they make an excellent refresher whenever I can't remember how this or that bit goes. The course itself was steady in its progress, presented by a mix of expert plant people, expert communicators and experienced guides. Through the sessions we had a lot of access to good people, and the same attentive listening skills I'd developed for radio came into play. This didn't stop at the end of the training course. The Guides expect that you'll need to continue educating yourself for as long as you are a guide, and provide good people as a regular feature of their ongoing meetings on the basis that such people are a reliable and accessible source of updates and deeper expert knowledge.

In my own time, there was a lot of extra reading. We each had our own copy of a book that listed, with illustrations, every plant in the park. Mine is covered almost wall to wall in annotations. I'd been in the habit of wandering through the local public library's garden and plant section whenever I was looking for something random to read. Now I went back to particular books and looked for extra details, especially those related to the texture, smell, or taste of a plant. And I wrote them down. I looked in other libraries too. Again, this was largely before the Web. The Web certainly existed, we'd used it a little at university, but home and business Internet accounts didn't have wide market penetration yet. When it came to looking for things online, you might find something useful, but it was a bit hit and miss. Sites I use regularly now either didn't exist at all or were in their very early stages with not a lot of their physical information yet online. And at that time, I don't think I even considered the Web as a useful source for information on botany and plants. It was great for some topics – but outside of the Web's obvious strengths a library was a far better bet.

One of the side effects of this that I've since regretted many times is that I didn't record references for a lot of my annotations. There were three reasons, I think. The first was that at least to begin with I remembered which books I'd read things in – there wasn't a plethora of them, after all. The second was that I was giving guided walks rather than writing a paper, so my audience really was all about the 'edutainment' rather than taking it seriously themselves. And the third was, and this seems very strange nowadays, I trusted my sources. Libraries were curators of quality books.

So were publishers. If a book had got to publication and then been selected for a library shelf, it was going to be on the level. And while botanical facts aren't immutable, especially in a state where we're several thousand species short of finding and naming all the plants we think are here and botany is a cutting-edge science, they're not that quick to change either. So a book is likely to be a reasonably permanent source of information.

Notes in a book aren't something you can stop and read when leading a walk. All the facts, the stories, have to be at your fingertips for that moment when someone asks a question. I learnt my notes by going on walks by myself, by seeing plants and looking them up, then touching or smelling them or scuffling around the site looking for proof of what I'd just read. Explaining to other bushwalkers why *I was down on my knees with my butt in the air poking at a plant only 2 cm high was tricky the first couple of times, but sometimes they'd get down on their knees to look at the specialised roots of a rare plant right along with me. And once down, I knew they'd remember it. You can forget words from a page so easily, but scrapes on the knees remind you they exist for the next couple of days. As a walk leader, I'd sometimes use a magnifying glass as a visual cue to encourage walkers to physically change their viewpoint, to get down or get close, to literally step through the screen they were seeing the world upon and stop treating it like pictures on a page. What they saw through the magnifying glass didn't matter. What mattered was that they moved.*

Moving into the future, I'm still learning. The way I manage my learning is quite different now. I still occasionally collect interesting books – I can even afford to buy my own. But I'm selective because there are so many more available. What's of most use to me is the Web, now laden with so many useful websites. Sure, there's a lot of dross. Especially if you want to know about the uses of plants – "natural medicine" misinformation abounds. My general science training kicks in here nicely, and I try to be thorough about vetting my sources, cross-checking information, and making sure the sources I'm cross-checking against aren't just plagiarising each other. I follow the rule of thumb of "Never trust Wikipedia until you've read the research". Also the rules "Pliny and Aristotle didn't always know what they were talking about", "just because someone gave two plants a similar common name doesn't mean you can conflate their chemical properties", and "if it sounds like a great story, someone might have made it up".

Then there's the websites that aren't meant to be objective and authoritative, part of the New Web. User forums, wiki-style maps, social knowledge collators. Like the site which aims to collect and collate information on what crops are being grown at what times of year in highly localised areas. You can look at other gardeners close to you and see what kinds of things they're growing, and whether it worked or not. Sites like the blog sites of environmental survival experts and teachers, who post about interesting plants they encounter as a way of engaging with the community they hope to teach. When I read these, I know that I'm getting anecdata rather than scientific data – but maybe if I want to grow lemons in the city I don't need research, I need to hear from someone who tried it and got it right. These sites can fail the "10% of your brain" test, in that they can propagate any popular error because people "feel it's right" and "know it's true" so happily repeat it. But they do allow a crowdsourcing of intelligence that – if sieved gently – can turn up all kinds of useful things.

The one thing I miss in using the Web is the physical. Nothing really substitutes, for me, for that getting down and into something. So I have a test garden, with several test beds, and a family that patiently puts up with me meddling with microclimates, sorting strange seeds, or cultivating compost whenever I have a moment. They also pretend they don't mind that the garage is full of reconnoitred rubbish for building garden contraptions. Some of my learning stays on the Web or in my head as a kind of wistful dreaming soon forgotten, but some of it leaps out of my fingers and into the dirt. And then I sit back and see how I did.

Before enrolling as a guide, Tiki was accustomed to volunteering and regarded it as part of her role as an active member of a community. Being a botanic gardens guide seemed a natural extension of her long-term interest in plants and her previous experience as a communicator, so she knew the benefits of listening well. The level of expertise demanded of her as a guide, however, came as a surprise and she had to *"to formalise and deepen a lot of the knowledge"* she already had. This led, in turn, to pursuing further knowledge on her own account. The help of expert mentors and the information found in books were significant in this regard. Like Liz, this process of learning about science has persisted, to the extent that Tiki has become an experimenter in her own garden.

Kristen's Story: In the Galleries

Kristen had a limited formal science background although, in her home life as a child, science played a valued role. She qualified with skills in business and hospitality, which led to a career in this area. Kristen's decision to enrol as an Explainer in an interactive science centre stemmed from her enjoyment in meeting and talking to people. She sought a deeper personal involvement in the world of others than the hospitality industry offered to her, and the demands of the Explainer role satisfied that goal.

My childhood was spent moving around a lot, both overseas and at home, as my father's job with various government departments, the WHO and the United Nations required moving every three years or so. He was very involved with health matters, safe water and food for example, so science was an important part of that. My schooling was quite disrupted – sometimes I was ahead in some subjects, other times behind. I did study science in school, with a year of physics and a year of chemistry in senior high school, but I don't remember much of that. I was more interested in my minor subjects, like photography. When I left school I did a Diploma in Business Administration, and another in Advanced Hospitality Management and some other certificate courses – I wanted to go into hospitality.

For quite a long time I worked in the hospitality industry, in restaurants and a casino, but seven years ago my sister got a job at our local science centre. As children, we had visited the science centre many times, my family were members and we loved it. My sister seemed so happy there that I applied myself just over three years ago and became an Explainer. We have to learn new things all the time, every day there is something new. The Explainers are a mixed group, some are expert scientists – for example, we have doctors, geologists...but even they have to learn when the topic is not familiar to them.

When I first joined, I spent the first few sessions with a "buddy". It was a different person each time, so I observed many different ways of explaining and talking to different audiences. They gave me tips on explaining, and helped with useful references: gallery notes (which explain all the exhibits in each gallery), magazine articles, videos, books. It is not enough to understand what an exhibit is about, you have to work out different ways to explain it for different ages and different people. Also adults sometimes ask how they might explain the idea to their children at home, or where they can find more information to further their own knowledge. It is great when children return to the science centre and share new knowledge they have discovered – you can learn from them too!

When you start, I found it helps to focus on the main galleries first to become very familiar with them, then move to the smaller ones There are additional resources available all the time – books and magazines in our tearoom, the Internet of course, and so on. Basic knowledge must be followed up with more in-depth information. Sometimes staff with specialist knowledge write up additional useful points, so we can learn more topics of conversation. There is such a large shared knowledge base and it is great to be able to tap into that.

Sometimes it is difficult to explain an exhibit – you can be asked difficult questions. I spend a lot of time doing more research because, when you are not a specialist, you need a broad base of knowledge to draw from. We carry intercom devices and, if I am really stuck, I can pass on my question and hope someone can provide the answer. I enjoy the confidence of feeling "I know this" but if not, I trust that someone will give me a quick answer.

Telling stories is very important – children especially respond to stories and being passionate and enthusiastic is critical in this job. I also think demonstrations are very important – little children need visuals to understand the world around them. For example, I love baking and when you are baking with children, it is not enough to say the cake should cool before you ice it. If they ask why this is so, let them try icing it while it's still hot. They soon get the idea! Hands on experiments help people remember, it has a lasting impact, and it helps to ask: What if? Why? What happened?

I think my people skills came from my experience in hospitality. I think I am good at reading people – knowing when to approach them and when they would like to be left alone to play. That part wasn't hard, but the explaining skills came from watching different buddies. I still do that, I will walk through a new gallery and listen to different Explainers and learn from them. My present job is more wide-ranging; I take on different roles but I still learn from the others. I myself have been a buddy too. I find passing on my knowledge whether it be to other staff or visitors to be very fulfilling.

We share a lot of our knowledge and interests– we have a staff science Facebook page started by a young Explainer, where staff post various scientific articles for us to look at. A particular Explainer, who has been a staff member for many years, writes the gallery notes for us – I have learned a lot from him.

From the start I had confidence that I could do this. I can relate to people, I just had to acquire more science based knowledge and I knew I could do that by researching and studying. We are constantly helped with regular all-staff training sessions, with updates, with one-on-one help. We always have access to experts when a new exhibition opens.

I think this place is about giving people the chance or choice to make up their own mind about what they would like to learn. The exhibits are all interactive and this lets people decide what they would like to engage with, this helps them learn and remember. They can choose what they are interested in, and that's how it should be.

It is a great job, with generous, sharing, passionate colleagues. There is nothing better than being able to say that I learnt something new today or I helped someone else learn something new. Having a job that allows you to learn is very rewarding, but then being able to explain things to other people gives you a purpose. I love it.

For Kristen the challenge was to learn science of sufficient depth and breadth to be able to answer a wide range of questions. Further, she had to find out how to modify her scientific explanations to suit audiences of different ages, from very small children to elderly adults. These audiences also have differing scientific knowledge, and this required being able to "read" their desired level of interaction. The task was made simpler by a tested method of training in the science centre which involves mentorship and modelling. Part of this mentorship takes the form of pointing out relevant resources. The culture of the group encourages development as an Explainer and, importantly, values in-depth knowledge of the topics to be explained. Colleagues remain a valued resource. Easy access to resources and experts facilitates this knowledge-seeking. Kristen's training is ongoing, as is her personal research, and she continues to find it rewarding.

Warren's Story: A Science of Place

From early childhood, Warren was an avid reader and had an interest in science, particularly technology. Perhaps not surprisingly, he opted for an electronics qualification and early work in radio. His interest in becoming an interpretive tour guide was sparked by a fascination with his local environment, which led to his joining local conservation groups and hence becoming a guide.

I was raised on a farm in Southern Hawkes Bay, New Zealand. It was a mixed farm – sheep and beef cattle being raised on the hilly part, and a medium-sized dairy herd on the flat. My father had limited schooling as he took over the farm after his father's early death. My father was always keen to apply the latest agricultural science to the farm, and was the first in the district to build a herring-bone style milking shed.

I recall a visit to an open day at Massey University (then Massey College) in Palmerston North when I was very young. Massey University specialised in Agricultural Science and I recall being fascinated by some of the demonstrations. I think I saw my first Bunsen burner!

My mother had a keen interest in all types of knowledge and we had an expensive 10-volume set of Arthur Mee's "The Children's Encyclopedia". [1] *This was favourite reading for me and as I suffered from asthma as a child I often had days away from school when I would read these wonderful books. I was drawn to the pages about technology (amazing cut-away diagrams of ocean liners and steam locomotives) and science (the edition I read was probably from 1950 but was modern enough that the age of Earth was placed at 3 billion years).*

My grandparents bought me a transistor radio with shortwave bands when I was about 11 years old (1964) and I discovered I could listen to the Voice of America broadcasts including the voices of astronauts as they orbited the Earth. I couldn't hear what they were saying but I recall being very excited that I could hear their voices at all! And yes, I recall exactly where I was when I first heard of the successful first landing on the Moon.

However, the biggest influence on my interest in science as a child was my Uncle Bert. Uncle Bert was a scientist, working in the Naval Research Laboratory at Devonport, Auckland. He would always answer my science questions and encouraged my interests in space exploration and electronics. He was a rational thinker and also liked to challenge orthodox ideas. I found out later that the Government of the day wouldn't allow him to import a computer for the Research Laboratory, so he imported the components over several shipments and built it himself anyway. He also had a spectacular model railway layout, so that showed me some more of the benefits of being a self-starter with good technology skills.

At secondary school I gravitated to science subjects with a much greater affinity towards physics and geography rather than biology and chemistry. My interest in electronics grew and I enrolled in night classes to study for a ham radio operator's license. Having achieved this, I took the unusual step of sitting "Electricity" as a School Certificate subject.

I was intending to carry on to University but it was the early 70's, I was bored with my final year of secondary school and when a position was advertised for a Technical Trainee with the New Zealand Broadcasting Corporation I jumped at the opportunity. The position was with the Auckland branch so I assumed I would be placed in the bright lights of the city of Auckland. Instead they placed me in a smaller town than I had been raised in. In simple terms I dropped out of the NZBC and wandered off to follow my hippy dreams for a decade.

By the early 80's I was living in Dunedin and had rediscovered the joy of reading. I borrowed heavily from the public library and read anything – so long as it was non-fiction. I recall reading an article in CoEvolution Quarterly "Where You At? A Bioregional Quiz?" I took the concept of really knowing about the place where I lived to heart and started to learn more and more about Dunedin, the Otago region, and southern New Zealand. I wanted to be very knowledgeable about my part of the world.

I joined a local conservation group (Save The Otago Peninsula, which later, in 1987, gave rise to the Yellow Eyed Penguin Trust). The Otago Peninsula was becoming known – locally and globally – as a stronghold of some very rare and interesting wildlife. Both the Royal Albatross and the Yellow Eyed Penguin were attracting visitors from around the world. My neighbour was developing an eco-tour business and asked if I would like to work as a tour guide. The tour content was a mix of human history and natural history. I found I had to learn a lot about the region's Maori and European history so I visited museums and read some books from the local public library. I found that it wasn't hard to get enough information to present a "basic" tour but I also found that visitors would ask questions where either their interest lay, or I had not fully explained the linkages between certain historic events. To close these gaps in my knowledge, I either quizzed local experts or found books that covered the events more fully.

For the natural history I found that, as well as filling the gaps as above as I developed my commentary over time, I had particular topics that I found interesting. For example I found that the area was based around an ancient volcano and it had previously been an island. I studied these geological facts by reading and visiting museums. Of even more interest to me were the flying skills of the Royal Albatross. (They can cover hundreds of kilometres over the ocean each day with little apparent effort.) The science of this was intriguing. Most natural history books covered the subject fairly lightly, but I found an article in a copy of Scientific American which explained the albatross's application of dynamic soaring in detail. I was very excited to learn this and made it one of the features of my commentary. When a guide is passionate about a subject, that enthusiasm comers through in his/her delivery and customers are also enthused and interested. I found I enjoyed the work immensely and since then tour guiding has been my main occupation.

Along the way I found that this type of communication also goes by the name "interpretation" – essentially interpreting science and/or history into a form that non-experts can easily understand. I helped form a networking organisation for interpreters – INNZ Interpretation Network of New Zealand.

I have also worked as a marketing manager for a planetarium/observatory; as a team leader in an aquarium; tutor for those wishing to join the tourism industry; and as an interpretation advisor within the Department of Conservation. These days I own a business in which I offer consultancy to tourism businesses in New Zealand and Australia.

A considerable amount of a tour commentary in nature-based tourism is focused on science, whether it is geology, botany, ecology, geography, agricultural science, anthropology, astronomy. To make the concepts and theories easily understood by those on the tour, many tour guides employ the interpretive approach which has four components. The interpretive theme is the "take home" message or tour concept. It is the structure on which the entire presentation is built, and relates to the specifics of the site or feature being talked about. To be effective the theme needs to capture the group's attention and stimulate their interest. Visitors need to be able to understand and connect to the feature being explained. Also, structuring a commentary around a theme also makes it easier for me because it helps me to select relevant information from the vast range of facts available.

The presentation needs to be organised into a clear, logical sequence or the whole tour comes across as a series of random, unconnected snippets of information which don't really enhance visitor enjoyment or learning. I need to make my interpretation relevant for every group of visitors, to connect with their previous knowledge and experiences. This also means avoiding jargon and complicated explanations, and instead using everyday examples, comparisons and analogies. My visitors are in recreation mode and don't need to pass a test at the end of the tour! I find that incorporating simple hands-on experiences that stimulate other senses such as touch, taste, and smell are effective and enjoyable.

I need to carry on a process of lifelong learning – constantly updating and improving my knowledge and guiding skills. This not only keeps me up to date with current events – whether local earthquakes or environmental issues in a guest's homeland – but keeps me interested and passionate about the subject matter.

I read a lot – still non-fiction – and get through 150–200 books each year. Nearly all the books are accessed through Auckland public library. About one third of those books are about the subjects covered on the range of tours I conduct regularly. I also have a current reading list for the guides I train and I use websites and Google when wanting quick updates. Then there are TV documentaries, museums, guide colleagues and, quite often, the guests themselves, as other sources of information.

The only regret I have is that it took me until around 2004 to discover the interpretive approach to tour guiding, and all the wonderful resources that have rolled out from interpretation organisations around the world. The amazing resources within the Dunedin, Waitakere, and Auckland public libraries have enabled me to pursue a very satisfying career.

"Libraries will get you through times of no money better than money will get you through times of no libraries." – Anne Herbert

Warren is very much a self-taught interpretive guide, although he now trains others in this profession. Unlike Tiki and Kristen, whose explaining was very focused on the interest and relevance of particular scientific objects – plants or interactive science exhibits – Warren has had to learn to integrate information from a range of disciplines and a range of sources. His knowledge has had to relate to the specific location of the tour visit: its flora and fauna, its legends, its history and the relationship between earth, sky, and water. His knowledge-seeking strategies have had to be wide-ranging and continuous and he reads a large number of library books, as well as using the Internet and television. Warren has a scholarly and informed approach to explanations and interpretations, and he has extended this approach into helping to found a network of interpretive guides across New Zealand.

The training which Tiki and Kristen experienced was typical of what explainers or "docents" are given in good public institutions of the kind described by Richardson (2012) in a study set in the San Francisco Exploratorium. In this study, she found that explainers learn science as part of their perception of their role. "The processes of learning science for one's self and engaging in the practice of teaching science to others go hand-in-hand for most Explainers" (p. 97). Her subjects integrated their personal learning into the practice of explaining, valuing the environment of learning generated by the program. In the program, they were given role modelling, access to expert content through visiting scientists and practical experience. The environment of learning was very supportive. Richardson added that the "community of practice" in which these explainers operated was very important.

Tiki and Kristen, too, were given written materials, role modelling, access to experts, and practical experience within a valued environment of learning. Warren, on the other hand, forged a new community of practice through his own experience, which was in many ways a harder and lonelier road to follow.

Bailey (2006) found that museum educators have to continuously learn on the job to keep up with their practice. Despite the very thorough nature of the training, both Tiki and Kristen found it necessary to supplement this by seeking more information from available sources. For Tiki, use of trusted books was very important, to the extent that she proceeded to explore the plants in the garden with books in hand. Today she uses the Internet extensively but still relies to some degree on authoritative books. New media are now included in her learning repertoire. Kristen has access to a range of books and other resources in the science centre, but still explores other sources. For Warren, the library has been a critical resource because his avid consumption of books needs to be supported by an ongoing supply. He consults other sources such as the Internet and watches television when relevant to ensure that his knowledge remains current.

Commentary on the Five Case Stories

It is useful to compare the roles of the people in this chapter to that of the "motivating instructor" outlined by Wlodkowski and Ginsberg's (2017). Such an instructor has empathy for the learner, understanding their needs and level of experience. The instructor's enthusiasm is an important factor, valuing what is being taught and explaining clearly. The critical issue here, however, is "expertise": The instructor must "know something beneficial" to the learner and "know it well". In different ways, the narrators in this chapter sought out the resources to convey expertise to them personally and, through them, to their audiences.

In the case of the explainers, empathy for the learners has to be achieved immediately, on the spot, and with strangers who are frequently adults. Understanding the audience's needs and their level of experience can only be achieved through careful presentation of the science and judicious questioning and sharing of knowledge. Good listening skills are essential, as Tiki soon found in her radio work. At the same time, the "audience" must be "entertained". "Explaining" is a very different skill from the teaching roles of Paul and Liz, whose audiences were children with specific problems to solve. As well as enthusiasm for the subject, the instructor must have the ability to explain clearly in an entertaining way and with relevance to the listeners, who themselves are likely to be a mixed group.

Although the audience for the science explanations has been very different for all the stories in this chapter, there are many parallels in the approach to explaining. For all our storytellers, their scientific explanations have been a "translation" of their own understanding. Successful explanations in science have distinctive characteristics. Gilbert, Boulter and Rutherford (1996a, 1996b) stated that they must be in a language that the recipient of the explanation can understand. They must satisfy the questioner, sometimes prompting further questions. They must be appropriate to the age and interests of the questioner, ideally encouraging further search for knowledge: they do not tell too much. These aspects have been summarised by Gilbert et al. (1996a) as follows: A good explanation is relatively simple (the condition known as "parsimony"), it is generalizable, and it is fruitful. The three explainers in this chapter have mentioned these aspects as part of their personal philosophy.

The stories in this chapter have highlighted the importance of context, in which concrete analogies are always helpful. It is surely no coincidence that all the stories in this chapter involve these aspects. Paul talked of visiting the museum to see model dinosaurs, while Liz made models of DNA and the Periodic Table. Tiki encouraged visitors to get down on their hands and knees:

> *You can forget words from a page so easily, but scrapes on the knees remind you they exist for the next couple of days… I'd sometimes use a magnifying glass as a visual cue to encourage walkers to physically change their viewpoint, to get down or get close, to literally step through the screen they were seeing the world upon and stop treating it like pictures on a page.*

Kristen thought stories and demonstrations were very important: *"Hands on experiments help people remember, and it helps to ask: What if? Why? What happened?"* Warren, too, knew the value of relevant experiences: *"using everyday examples, comparisons and analogies … I find that incorporating simple hands-on experiences which stimulate other senses such as touch, taste and smell are effective and enjoyable."*

For all our case story authors, conclusions about the value of these elements of explanation were reached through experience and trial and error, not through prior knowledge. It may be argued that Paul knew about the museum because of his own qualifications, but he had to make a judgement about its value to Emma when other sources failed to satisfy her: it was now time for the concrete experience. In Liz's case, she had found that the analogical models had helped her own learning, and she extended that idea to her children. The three explainers seem to have arrived at similar conclusions through role modelling, careful observation, and personal reflective learning.

Like the parents in this chapter, the three explainers were highly motivated and had confidence in their own ability to take on the challenge of finding information and interpreting it for others. All of our authors had trust in their sources of information, which particularly relates here to trust in printed materials such as books. All had access to expert knowledge, either in the person of local scientists or through museums. For all of them, the language of science had to be modified for various ages and cultures. This requires flexibility and empathy for the audience. All the explainers mentioned that an integral part of their ability to explain well is enthusiasm and passion for the job. For the parents, enjoyment in the interaction with the children was evident and, for Liz, subsequent personal learning was fun and challenging. It seems likely, therefore, that personal enjoyment and satisfaction are important factors in this kind of learning.

The patterns of learning science described in this chapter have differences, but many parallels. These will be further explored and analysed in subsequent chapters of this book.

Note

1 *The Children's Encyclopædia* by Arthur Mee was published in London from 1908 to 1964. It was published in the United States as *The Book of Knowledge* (1910).

References

Bailey, E. B. (2006). Researching museum educators' perceptions of their roles, identity and practice. *Journal of Museum Education*, 31, 175–197.

Bandura, A., Barbarinelli, C., Caprara, G. V., & Pastorelli, C. (1996). Multi-faceted impact of self-efficacy beliefs on academic functioning. *Child Development*, 67, 1206–1222.

Gilbert, J. K., Boulter, C., & Rutherford, M. (1996a). Models in explanations, Part 1: Horses for courses? *International Journal of Science Education*, 20, 83–97.

Gilbert, J. K., Boulter, C., & Rutherford, M. (1996b). Models in explanations, Part 2: Whose voice? Whose ears? *International Journal of Science Education*, 20, 187–203.

Gottfried, A. E., Fleming, J., & Gottfried, A. W. (1998). Role of cognitively stimulating home environment in children's academic intrinsic motivation: A longitudinal study. *Child Development*, 69, 1448–1460.

Juang, L. P., & Silbereisen, R. K. (2002). The relationship between adolescent academic capability beliefs, parenting and school grades. *Journal of Adolescence*, 25, 3–18.

Richardson, A. E. (2012). *Explainers' development of science-learner identities through participation in a community of practice.* Antioch University: Dissertations & Theses. 77. Retrieved from https://aura.antioch.edu/etds/77

Stocklmayer, S. M., Durant, I., & Cerini, B. (2011). Giving mothers a voice: Towards home involvement in high school science. *International Journal of Science Education, Part B*, 1, 23–46.

Wlodkowski, R. J., & Ginsberg, M. B. (2017). *Enhancing adult motivation to learn: A comprehensive guide for teaching all adults* (4th ed.). San Francisco, CA: Jossey-Bass.

6

LEARNING FOR WORK

Many people learn science because it is required for their employment, or because they wish to change their career to a more science-oriented situation. The need to learn "on the job" has been the subject of investigation by several researchers. In particular, the work of Michael Eraut (see, for example, Eraut, 2000, 2004a, 2004b) has encapsulated some of the enabling factors about a place of employment that help people to progress with knowledge and experience. Eraut (2004a) made the comment that "apart from being under-researched, the workplace context brings new perspectives to research on learning because it encompasses a wide range of more or less structured environments, which are only rarely structured with learning in mind" (p. 248).

Eraut (2000) defined learning as "the process by which knowledge is acquired. It also occurs when existing knowledge is used in a new context or in a new combination" (p. 114). In this paper Eraut discussed the term "non-formal learning" and the importance of "tacit knowledge". Non-formal learning in his typology includes three types of learning: "implicit learning, reactive on-the-spot learning and deliberative learning" (p. 113). All these terms are, of course, relevant to learning in the workplace and are evident in the cases we describe here, but the confusion between the use of "non-formal learning" in educational research and in the work of Eraut means that we will not use that term in this chapter. We will, however, examine the learning that occurred in terms of his three components, and tacit learning.

Eraut (2000) also distinguished between codified knowledge (that which is "officially" required to carry out employment duties satisfactorily) and personal knowledge. Codified knowledge is explicit and may be learned, perhaps through taking a course. This knowledge is then, however, incorporated into personal knowledge that is "affected by the learning context, so that subsequent use of that

DOI: 10.4324/9780429434303-6

knowledge in a different context will require further learning" (p. 114). This perspective is exemplified in our first story, where the need for further progress at work led to taking relevant courses which led to new directions – an employment pathway with which many readers will be familiar.

Learning In and For the Workplace

In this chapter, three very different career pathways are described, each of which required learning more science as an adult to effect employment changes. This did not necessarily mean a change in the fundamental nature of the employment, but led to further opening of doors and a greater interest in and understanding of the science.

Hugh's Story: An Experience of Life-Long Learning

Hugh has spent his whole working life in the aeronautics industry in England, beginning with an appointment as an apprentice. He completed middle high school with good grades across a range of subjects, particularly in mathematics, but had no further formal education in science or mathematics outside what was needed for his apprenticeship aviation training. This is his story.

> *After leaving school in 1988 with 9 GCSE's (Middle High School leaving examination) at grades "C" and above, and with an "O" Level in mathematics which I took a year early, I was accepted on the British Airways (BA) Apprenticeship Scheme and hence started the daily drive from my home in Woking to Heathrow Airport.*
>
> *Our first apprenticeship year was split between the BA Engineering Training Centre at Heathrow where we learned hand skills along with fortnightly block release to Southall College of Technology to learn aviation engineering and aircraft technology theory. Having been interested in everything mechanical as a child, along with being a keen aero modeller, this all came easy to me. During our second year we were released into the hangars to work on the aircraft undergoing heavy maintenance while continuing with our college studies, and finally during our third year we were placed on the "Line", working on the live aircraft as they arrived from passenger flights.*
>
> *After completing our three-year Apprenticeship, we graduated as Engineering Technicians, each of us with an Ordinary National Certificate in Aeronautical Engineering and a City and Guilds in Aircraft Engineering Techniques. [These are United Kingdom Technical Education awards.] At this time, aged 20, it dawned on me that I wanted to achieve more and progress further than being a technician. BA was very supportive in furthering the careers of those who would push themselves and hence I was able to obtain approval for further day release to Brooklands Technical College where I gained a Higher National Certificate (HNC) in Aeronautical Engineering,*

which satisfied my academic desire, and put me on a level with my peers who had gone on to do upper school examinations.

At the same time as studying for my HNC I also started the process of studying for (and subsequently obtaining) the various modules required to apply for my Engine and Airframe License, a qualification issued by the Civil Aviation Authority, which would allow me to certify an aircraft as airworthy and fit for flight. My License studies proved very time consuming and required a significant amount of effort and perseverance to complete, especially as at the time I was still finishing my HNC and holding down a full time job as an Aircraft Technician at Heathrow. However, working shifts did allow me some spare time to bury my head in my books (in between renovating an old house I had just bought). By 23 I was a Civil Aviation Authority Licensed Aircraft Engineer, and had completed my first Type rating on Boeing 747–400 aircraft, meaning I joined the ranks of the certifying staff with the responsibility of ensuring BA's fleet of "Jumbo Jets" were fit for flight.

The next step was to become a Supervisor and this was where a little luck came into play, as BA decided to replace their aging Douglas DC10 fleet which was operating out of London's Gatwick Airport, with B747–400's; however most of the "older" Supervisors working at Gatwick were not keen to spend up to four months in a classroom learning about a new (and more modern) aircraft type. Hence I opted to take up a position as a Supervisor at Gatwick's North Terminal running a small team of six or seven Technicians. Subsequently with the introduction of the Boeing 777–200 into the Gatwick fleet, I was one of the first to complete B777 training and add this Type Rating to my License. At the time, with only four of us approved on this new jet, we were very busy, not only at main base but carrying out engineering cover on crew training flights at various locations in Europe.

The latter role whetted my appetite to join BA's Overseas Pool of engineers (known as OSEs), who worked in various outstations to cover for those local staff who were on holiday, training courses, etc. This role was a further promotion to Technical Management grade, and by 25 I was one of BA's youngest OSEs. Along with being stationed in numerous locations including Rome, Madrid, Islamabad, Abu Dhabi, Lagos, Abidjan, and Paris to name but a few, I was selected for security clearance to allow me to become a "flying Engineer" to cover VIP flights. This culminated in me being involved with several high-profile trips with (Prime Minister) Tony Blair and Prince Charles.

It became apparent that life as an OSE, although fun and interesting, was not conducive to a happy and successful marriage, and hence in late 1999 I returned to Heathrow where, after a short period as a Technical Instructor on Boeing 767's, I ended up working in BA's Casualty Maintenance department where I ran a small team repairing damaged aircraft for both BA and other carriers, in the United Kingdom and overseas. It was during one such large repair that I met an Aviation Surveyor who was carrying out a damage assessment on behalf of BA's Insurers. It was at this time when I realised that I needed a new challenge and, as luck would have it, through my BA career I had gained great experience in not only technical skills, but interpersonal skills and was able to relate with everyone from the shift storeman to the Senior Management Team and was happy to address them all in my capacity as a Licensed Engineer.

Needless to say, with some trepidation I left BA after 16 years and joined Airclaims which was an Aircraft Surveying and Adjusting company within the Aviation Insurance industry and became an Aviation Surveyor. Out went the overalls, grease, and tools and in came the suits, shoe polish, and briefcase!

It soon became apparent that there was much to learn. The role involved giving technical assistance to the Aviation Insurance Industry, assessing damage to aircraft engines and spares and finding repair solutions, project managing repairs, and making financial recommendations to Insurers. Furthermore, the role also involves investigating and reporting on major losses and dealing with recovery and crash site clean-up, which in the case of a large airliner crash site can be a very sobering experience indeed.

Specialist accident site training was received from the Air Accident Investigation Branch (to allow access to accident sites) and further Air Accident Investigation training at Cranfield University was undertaken to broaden my knowledge in the investigative aspects of the role. The fore-mentioned and the technical aspects of the job were relatively straight forward coming from my background; however, an understanding of the Insurance industry was also required, which proved more difficult to grasp. Relationships between all the interested parties including Underwriters, Brokers, the Operators/Airlines, Lawyers, Surveyors, and the relevant specialists all need to be learned and it takes a long time to build up a good reputation in this line of work; however, it can take only one mistake or incorrectly reported event to ruin it.

Guidance and basic training was given by my employers, however I took it upon myself to read up on Insurance Industry rules, regulations and protocols from various publications and training documents handed down to me from others. As my experience increased, I was deployed on more and more complicated losses, initially being overseen by a Senior Surveyor but ultimately out in the field on my own. After approximately six years I was promoted to Senior Surveyor myself and subsequently assisted and help train those had recently joined our company.

Subsequently I was approached to join a newly formed Aviation Surveying and Adjusting team at Crawford where I am currently part of a small team located in our London office dealing with all kinds of accidents and incidents on behalf of Insurers. Typically the losses we deal with cover a wide range of sectors within the aviation industry including drones, gliders, microlights, general aviation aircraft, airliners, and even pre-launch satellites. I personally think I am privileged enough to have had an extremely interesting career, culminating in what I (as a Licensed Aircraft Engineer) would consider to be one of the most interesting, fulfilling and satisfying roles within the Aviation Industry.

I was fortunate to be mentored by some influential professionals within the Aviation Insurance field who themselves had built up great reputations over many years. This subsequently made life much easier for me at the time, as there are no specific Aviation Insurance Surveyors training courses or qualifications due to the limited number of people leaving aircraft engineering and switching to Insurance. Nowadays new recruits are encouraged to undertake generic Insurance training with certainly helps in understanding the Insurance Industry more quickly; however, nothing beats experience in the field which can only be gained by getting out there under the guidance of a more experienced Surveyor.

I look back at my career path and feel that the decision I made after completing my initial training to continue to learn and better myself was the turning point. At the time many of my peers decided that after all those formative years spent learning, they would stop and take some time out. Some of those I still stay in touch with are happy with the way things ended up for them; however others wish they had continued to push themselves and develop further, but now feel it is too late.

I realised then as I still do now, that continued learning and increasing one's knowledge opens so many doors, not only in business, but in personal life. Currently I'm still learning new skills, this time in sound engineering and band technical support. Whilst nothing to do with my career, these further skills have allowed me to help my children with their band, which has subsequently become a hobby for me and a passion of the whole family...and it helps make me feel younger!

Hugh's interest in aircraft began as a child and continued into adult life through his modelling hobby. Translating this into a career was, at first, quite straightforward with his acceptance into a well-established apprenticeship scheme. From this point, however, much depended on Hugh's drive to learn more about aircraft technology and, from the start, his pursuit of further formal qualifications that were organised and sponsored through the apprenticeship scheme. As he himself says, he could well have stopped there, since formal support did not continue past the higher certificates he had earned in a linear progression from his enrolment as an apprentice.

The first step outside this support was to study for his Engine and Airframe License. To pursue further study while remaining in full-time employment requires dedication and commitment, particularly when the content is technical and demanding. It is a lonely path. Hugh was rewarded with more career options and with promotion at a relatively young age. His subsequent move to become an Aviation Surveyor involved not only new technological understandings but an ability to integrate those understandings across other disciplines including insurance assessment and law. Enrolling for further formal university education to support this new career once again made demands on Hugh's time and commitment but his willingness to accept further challenges is evident throughout his story.

It is clear from Hugh's story that he was a reflective practitioner, constantly evaluating his current situation and possible choices. He saw opportunities and was not afraid to grasp them, even when some risk was incurred. His highly developed people-skills facilitated change and led to his acquiring useful and valued mentors for the various roles he describes. It was inevitable therefore that, during this career progression, he would also mentor others, something many of our authors were able to do as a result of their own learning.

Ketan's Story: Understanding Controversy

Like Hugh, Ketan was interested in technology from a young age, and this led to a fascination with science fiction. Although he obtained a degree in microbiology, his subsequent career required very different science knowledge. He currently works for an energy company, assimilating and disseminating information about science.

> *My passion for science was forged within the walls of a decommissioned space shuttle. I think I was around six when my family visited the Kennedy Space Centre. The scale and power of the machinery and technology was inspiring, but thanks to a regular diet of "Star Trek: The Next Generation" and the sci-fi comedy like "Red Dwarf" I also knew that the act of sending humans into space was closely tied to scientific endeavour.*
>
> *These close encounters with spacecraft provoked curiosity, but it was through works of fiction that I truly began to understand science. As a child I had a love for dinosaurs (fuelled through avid consumption of an expensive 'collect-the-whole-set' magazine series), and so, when I first heard about Michael Crichton's Jurassic Park I refused the junior novelisation and dug straight into the jargon-heavy adult-version. Unlike Star Trek and Red Dwarf, this wasn't the stuff of starships and exploration. It was a cynical, violent warning about the dangers of unbridled scientific power. As Crichton wrote in Jurassic Park, "Science can make a nuclear reactor, but it cannot tell us not to build it. Science can make pesticide, but cannot tell us not to use it. And our world starts to seem polluted in fundamental ways – air, and water, and land – because of ungovernable science."*
>
> *Crichton's warnings of scientific overreach (framed in bouts of gory drama) only furthered my enthusiasm for science and technology. It established, in my mind, that technological expansion through science ought to be a cautious equilibrium, not a blind push for progress. My enthusiasm for science and technology was boosted in my mid-teens: seeing the label "science" erroneously applied to things that weren't scientific filled me with frustration. While my friends nervously sent text messages to girls on their Nokia 510s, I spent my high school lunch breaks scanning the pages of Richard Dawkins' "The Selfish Gene". I was genuinely captivated by the elegance of Darwinian explanations, and simultaneously enraged by those who casually rejected them.*
>
> *When I started university I chose molecular biotechnology as my degree, and, with some second-year hiccups in personal discipline leading to some failed subjects, progressed through my University's undergraduate system. It wasn't until I stumbled upon the neuroscience major offered at the University that I became truly invested in my studies. Combined with psychology, the degree offered an answer to something that had been bugging me for a long time: why do people reject science, and believe weird things instead? My marks rose instantly – I completed assignments with ease, and attended every single class, because every single piece of information fascinated me.*

Throughout my later years at university, I moved away from science fiction. I read about why people are pushed to the edge of belief, and why we so readily reject science. I managed to snag a copy of a book called "Why People Believe Weird Things" by Michael Shermer. This moved past the literary snark of Christopher Hitchens' "God Is Not Great" and towards a more compassionate investigation of the neurological reasons for belief. I discovered that although some beliefs might be weird, the act of believing them was quite normal. I was intrigued by Daniel Dennett's explanation that humans have a soul, but it's composed of tiny machines. I discovered Carl Sagan's "Demon Haunted World". Sagan's writing on space and exploration pinged those instincts formed in my childhood, but his compassionate approach to scepticism, and his systematic explanations of pseudoscientific belief were thrilling in a completely novel way.

After I finished my undergraduate degree I was ready to explore neuroscience further. I had an Honours thesis in mind, but I stalled and ended up working as a research assistant between other terrible jobs, including a long stint in a call centre. I'd write blog posts that sarcastically critiqued homeopathy and creationism and share them on Facebook. I also consumed every single piece of pop-psychology I could. By this stage I'd essentially given up on long-form literature – long drives were accompanied by endless talks and podcasts exploring religion, science, technology, and psychology. This was my scientific training – I cut my teeth on a range of weird, unbelievable pseudosciences. I learnt the value of understanding the details, and the varieties of the scientific method. But I also learnt how to understand and sympathise with those who held to their beliefs in the face of contradictory evidence.

After a year, I realised I had to leave my terrible job and seek out something more aligned to my passions. I searched for jobs suited for a science graduate and had some truly terrible interviews. I came dangerously close to accepting an offer to be one of those sales reps who go around convincing doctors to buy pharmaceuticals. Eventually I applied for a job helping to establish a remote analytics and monitoring centre at a little-known energy company. I started out performing complicated analytics – I had to recall my statistics classes (which meant digging out my textbooks once more), and I learnt about the mechanics of the electricity market, airflow around wind turbine blades, and electrical engineering. My equally intimidated newly-hired colleague and I applied ourselves to the challenge, and successfully set up the 24-hour monitoring centre which is now in place. The centre monitors a fleet of wind farms into the country's two biggest distributed electricity networks. We worked a range of night and day shifts and were often on-call. We spent hours buried in complex, enormous spreadsheets.

But I soon discovered that the world of renewable energy was linked to my interests around scepticism, the scientific method, and psychology in more ways than one. Wind power has been a regular target in the media, often targeted by local opponents and climate change sceptics. My university days prepared me for lengthy periods of analytical investigation, but it was my consumption of books, TV, talks, and articles that equipped me to deal with two major subduction zones between hard science and the human brain – the space where the outcomes of science and the reactions of the human brain collide violently. Those denying climate science were dominating the national news, and sometimes-overlapping groups insisting that wind turbines were causing a mysterious new disease dominated the local news around projects. I wouldn't have been equipped to deal with either phenomenon without a good understanding of the creation and propagation of pseudoscience.

The issue of alleged wind farm health impacts was immediately familiar to me. It bore the hallmarks of every pseudoscience I'd encountered. A heavy reliance on anecdotal evidence, experts with inflated authority but little scientific experience, the reversal of the burden of proof ("wind companies have to prove wind turbines are safe") and an abundance of truly improbable claims all conjured up comparisons to my experience of this phenomenon with other technologies: vaccination, fluoridation, Wi-Fi, smart meters, mobile phones, high voltage transmission lines, cell towers...all these technologies bear their own form of what had been dubbed "wind turbine syndrome". And, sadly, reactions to these phenomena have always been left wanting, with regards to the amount of compassion and empathy shown by scientific authorities and industry figures.

Of course, my first instinct was to debunk the theories. As with other pseudosciences, the claims around wind turbines involve a range of falsehoods. Biological and acoustic figures were fudged to sound scary, and scientific research was cherry-picked or poorly conducted, all with the aim of simulating a public health disaster. However, I also soon realised that there were good scientific reasons for the prevalence of the "wind turbine syndrome" phenomenon. People who misperceived the risk of a technology would often express their fears through a narrative of helplessness and disenfranchisement. In countries where communities were deeply involved with wind projects, wind turbine syndrome was unheard of. This was the Saganist in me, telling me that, although the rejection of science may be undesirable, it happens for a reason.

After dedicating much of my spare time to writing and researching the issue, I finally moved from the now-mature operations centre into a research and communications role at the same company. It's from this vantage point that I dived straight into the climate debate – another contentious nexus of passionate public debate. The systematic, widespread rejection of climate science is a social phenomenon that has now been well-studied and can be understood in the context of political worldview. Those who adhere to a libertarian, individualistic view of economics and politics are horrified by the idea that power plants with high emissions ought to be regulated for the greater good. This manifests not only as an attack on regulatory policy but, fascinatingly, as a direct attack on the science of climate change.

My own curiosity is focused on those who simultaneously claim the evidence of climate change is too weak, and also that the evidence for wind turbine syndrome is inarguably strong (despite scientific authorities stating precisely the opposite). There is a scientific explanation for this emergent dichotomy. Though my university education trained me to critically examine the numbers, a range of science communicators taught me to dig deeper into why these groups have seemingly placed blinkers on their own vision.

My day job involves the digestion and dissemination of scientific information. Sometimes, my prolific social media and blogging commentary serves a therapeutic function – a sarcastic appeasement for those of us growing weary of the lack of science in political leadership. But when it comes to finding ethical, analytically honest solutions to the roadblocks facing clean technology, the solutions lie in understanding the motivations of those who react to widespread technological change. These are the tangible consequences of a more compassionate approach to scepticism, tempered, of course, by a willingness to always call out falsehoods, where it's appropriate and useful.

Elon Musk, the billionaire genius behind the gargantuan companies Solar City and Tesla that are revolutionising energy generation and consumption, also owns SpaceX – a successful spacefaring venture which aims to colonise Mars. It's funny how so much of what I do with clean technology links back to that starry-eyed, unbridled enthusiasm I had instilled in me when I was younger. The wonder of space travel driven by my early love for science fiction still drives much of my passion. I go out of my way to watch spacecraft launches. At weird hours of the night, I sit transfixed to the live feed of spaceships on launch pads, with coolant hissing in the background. I get to experience science on a wide spectrum – ranging from the starry-eyed wonder of contemporary space travel, through the mechanics of the human mind, and the mathematics of clean technology's integration in our energy system. It's great, and I have a collection of skilled communicators to thank for the inexorable growth of this broad spread of scientific wonder.

It is clear from Ketan's story that his heart has always been in controversial science, dating from his early encounters with science fiction and progressing to current global issues confronting many (especially Western) countries. Even while completing his degree, he found himself drawn to scientific controversy and the need to understand alternative beliefs. Significantly, he "stalled" at the prospect of furthering formal education beyond his bachelor's degree and, instead, found himself seeking another pathway. It seems that his initial application to the energy company was almost accidental. Nevertheless, it led to revision of long-forgotten statistics and to on-task learning about energy supply and distribution, a challenge which he met with enthusiasm and commitment. At this point, his interests in neuroscience, in alternative beliefs, in science fiction, and in issues around clean energy fused into a role for writing about science.

Ketan's career is very different from Hugh's highly focused and determined pathway. Nevertheless it is possible to discern driving influences from the outset, in his interest in technology and his fascination with science writing. He has not recorded a presence of strong mentors – instead, his role models have been great science communicators who have motivated him to commit his own ideas to blogs and social media. In terms of learning more science, he also has gained much stronger insights into the issues of energy that are the focus of his current work. Looking back, he can see that early influences are still present in his worldviews and his current occupation.

Keith's Story: Life Is a Garden

The last story in this chapter has echoes of both Hugh and Ketan's experiences. Keith, too, had childhood interests which translated in later life to his work. He, however, made a conscious decision in mid-career to change to a completely different occupation.

I grew up in Sydney, on a three-quarter acre bush block adjacent to Garigal National Park. The national park was my playground, exploring the bush, its flora and fauna. I was able to catch yabbies (fresh-water crays) and observe native birds such as parrots, lorikeets, and honeyeaters. Over the years I occasionally witnessed wallabies and lyrebirds.

My parents were high school teachers, Mum a Home Economics teacher, Dad Industrial Arts. Both areas were creative endeavours requiring high levels of technical and practical skills.

My love of Australian plants can be traced back to our Australian native bush garden, and ideas of self-sufficiency probably can be traced back to our large 'veggie' garden, which provided most of our vegetables. I have strong and fond memories of visiting my grandparents in the Hunter Valley and watching my grandfather in his large vegetable garden and home orchard. Growing up, my hobbies included gardening, sailing, bushwalking, and camping.

After leaving school after completing year 10, I commenced a career in finance due to strong school certificate results in maths. I started in the banking industry before working in businesses, in civil engineering, public accounting, and in the seafood, agriculture, manufacturing and wine industries.

In my early twenties, following a move to South Australia, I enjoyed working for a fishing company dealing with tuna and the seafood market, and supplying prawns to Japan. I also worked for a small agricultural business dealing in jojoba oil, turning it into a commercially viable proposition.

I was never a traditional finance person, the numbers by themselves did not seem that exciting to me; however, looking behind the numbers did mean something. During my work with manufacturing I started to understand whole of business, sales, purchases, warehousing, manufacturing processes, and efficiencies.

In the 1990s, I went to work for a major wine company in their finance department for a few years before accepting a position in the viticulture and winemaking department to improve administration of their grape purchasing. This was because the company was expanding, along with the whole Australian Wine Industry. Here I received a total overview of the grape growing, grape purchasing, grape harvesting, grape crushing and winemaking processes, working with viticulturists, winemakers, winery managers, vineyard managers and independent growers, and with company computer programmers.

This was an exciting time, growing in excess of 10% per year, year after year. Each year I improved and developed the administration systems. I needed to know how it all worked, later on understanding the concept "from paddock to plate". This meant continued development of our grower grape-purchasing database, included standardised information such as grape varietal clones, soil types, and irrigation details, and harvest information (grape maturity testing, grape quality, grape disease details). Also, grape and wine research projects were starting to be incorporated into the database. I started to appreciate the process and rigour required in setting up, running and analysing results and finally publishing the results. I worked with highly skilled people and industry leaders. Several had PhDs or were studying towards PhDs looking at irrigation, grape maturity testing using the infrared spectrum, and grape disease.

My final years in the industry were spent in the vineyards and field offices purchasing grapes, which gave me first-hand experience and knowledge of planting vineyards, soil types, establishing irritation, grafting vines, disease control, control of yields, impact of climate, and monitoring each season's growing and harvest cycle.

I left the industry following a downturn during the early 2000s, and returned to administration. I did not enjoy going back to office work staring at computer screens all day. I missed the outdoors, discussing how to grow grapes, helping people record research, observing the seasons and its impact on grapes and wines produced.

In a bid to find a more enjoyable and meaningful work-life, what were my options? I referred to a career consultant, and after several tests and information sessions, the job suggested to which I was most suited was a landscape architect, which I dismissed due to the years of formal study required. However, the idea was planted, which I went back to over the next few years as I persisted with administration work, until I considered working in the horticulture industry. It would combine many of the things I enjoyed in life, the outdoors, helping people, growing things, observing the seasons. So I commenced running a gardening and lawn mowing business, which I believe incorporates the best bits from my previous work. It has been a steep learning curve and I feel there is so much to learn.

When I have a question in my current work I don't usually consult the web – I have a horticulturalist I speak to when I have a problem. There is a logical approach to solving problems. For example, I helped establish a native garden but there was poor growth and lack of nutrients – it turned out to be root rot from overwatering. How do you get to know that? I am not sure, but I go to my adviser and tell him what I have observed. You become a sponge in a rich environment.

I do have reference books, specific local ones put out by the horticultural society and they are in the car all the time. I am looking at further study, though, because there are gaps. I want to know more – I want to know why things happen. I might use the Internet to find a particular plant, but not much.

My Mum and Dad were responsible for encouraging the powers of observation and logic. Dad taught me to look and observe, to pay attention to detail, even though this was in the context of industrial arts (woodwork, metalwork, and fiberglass). Mum taught me how to observe nature while working in our native bush garden and seeing the interaction of plants, climate, and local birds and animals. Mum often talked about reaching and inspiring the kids at school, to make a difference. She was given the opportunity to co-author a new home science textbook, following criticising the existing texts on Home science, they did not cover the whole syllabus. Writing the textbook changed the family dynamics. Dad supported her, not only with research and proof reading but with helping with domestic chores. He asked how could we do this logically, what were the steps. They were a great example of sharing the work load, working as a team to achieve a goal. My godfather, who was also a school teacher, believed in inspiring and developing children with the classics, the arts and sport, giving them a hunger for knowledge; he believed in fostering their talents.

The powers of observation and application were further enhanced experiencing the scientific process at the wine company. Working with people doing research and completing PhDs, it was a rich environment to learn and develop.

Looking back, I would never have thought of me as, say, a doctor – school was too regimented and I wanted to get out and work. The financial world offered me stability, however for me it was the wine company, working with talented people striving to improve, to leave a legacy within the company and make a contribution to the industry as a whole, that changed my work perspective. They were enthusiastic, great communicators, they enjoyed what they did and did it well. My current horticulture career appears to be bringing together my experiences from my upbringing, my hobbies, and my best years in the wine industry.

Keith's experience of gardening and "the bush" as a child was deeply bound up with his family life. The national park on the boundary provided for experiences which laid a foundation of affection for the outdoors. His parents, however, also set an example, not only of involvement in education but concern for its effectiveness, and both parents taught him skills of great use later in life. They, together with his godfather and grandfather, were his first role models. Nevertheless, Keith did not stay in school past middle high school because he disliked the regimentation of formal education. It is possible to conjecture, even at this point, that Keith would end up in a less regulated work environment.

In terms of his scientific education, however, he had little background to call upon. Translation from school into the world of finance meant a focus on accounting and on people skills that made no call on scientific knowledge whatsoever. The various companies for whom he subsequently worked had agricultural interests, echoing his outdoor experiences and his hobbies, but his innate interest in plants was clearly brought to the fore when working for the wine company. Here he encountered mentors who were experts in the field and Keith learned enthusiastically not only about the cultivation of vines but about botany more generally. Clearly, he was a "sponge" even at this point, absorbing information in a manner close to what Eraut (2000) termed "tacit" or perhaps "implicit" knowledge.

Keith attributes much of his current approach to his work to the examples of mentors throughout his life. He has a horticulturalist whom he consults when necessary, but also uses specialised books. He is, overall, unable to say exactly where his current wide knowledge of plants has come from, but his approach to problem-solving reflects a willingness to commit to lifelong learning in the field.

Commentary on the Three Case Stories

It is evident from the three accounts presented here that early experiences had an enduring effect on subsequent professional pathways in science. Eraut (2000) referred to this kind of learning as implicit – it links past memories with current experience, and can influence future behaviour (p. 116). Eraut cited Reber (1993) who defined implicit learning as "the acquisition of knowledge independently of conscious attempts to learn and in the absence of explicit knowledge of what was learned"; there is no overt intention to learn and no awareness of learning at the time it takes place. Much of Keith's learning at the wine company was implicit; his learning was spontaneous and unplanned up to the point that he decided to become a gardener. The decision to change to this profession was, however, highly deliberative and his subsequent science learning had recognizable patterns discussed in more detail below. Reactive learning, on the other hand, is near-spontaneous and unplanned, the learner is aware of it but the level of intentionality will vary and often be debatable (Eraut, 2000, p. 115). Ketan was somewhat reactive in his learning, integrating past knowledge and interests into his current role at the energy company as the need arose, to gain insights into issues which spilled over into the public sphere.

The dominant mode for Hugh in his career path, however, was deliberative learning (Eraut, 2000, p. 116), in which he constantly reviewed and reflected upon his experiences and set new learning goals. These led, in several instances, to enrolling in formal courses for which the body of knowledge was defined and, subsequently, assessed. Even here, however, Eraut (2004b) made the point that the transfer of knowledge from education to workplace settings is much more complex than commonly perceived.

All three cases corresponded to an example quoted by Eraut (2000) from an earlier study, in which most participants formed some idea of their desired outcome, but took advantage of learning opportunities as they arose. In such a case, the intent is deliberative but opportunities are reactive. Eraut later added

> We also anticipate that at any one stage in a person's career, there will be both a group of learning trajectories along which they are explicitly and intentionally progressing and another group along which they are implicitly and unintentionally progressing; and that the composition of these groups will change over time.
>
> *(Eraut, 2004a, p. 266)*

Throughout these narratives, the concept of self-efficacy (Bandura, 1982; Bandura, Barbarinelli, Caprara, & Pastorelli, 1996) stands out as an important factor in achieving goals. Although doubts and fears were expressed, they did not deter Hugh, Ketan, or Keith from taking decisive steps that had long term consequences. In Hugh's case:

> *Needless to say, with some trepidation I left BA after 16 years and joined Airclaims which was an Aircraft Surveying and Adjusting company within the Aviation Insurance industry and became an Aviation Surveyor.*

Hugh accepted the risk, having enough self-belief to make the move. For Ketan, the self-belief was evident at an earlier stage:

> *After a year, I realised I had to leave my terrible job and seek out something more aligned to my passions. I searched for jobs suited for a science graduate and had some truly terrible interviews... Eventually I applied for a job helping to establish a remote analytics and monitoring centre at a little-known energy company... My equally intimidated newly-hired colleague and I applied ourselves to the challenge, and successfully set up the 24-hour monitoring centre which is now in place. ... After dedicating much of my spare time to writing and researching the issue, I finally moved from the now-mature operations centre into a research and communications role at the same company.*

Keith took some time to take the step of branching out on his own, a reflective process that gradually grew into a decision. Clearly, however, he had enough confidence in his own ability and knowledge to take this risk:

> *The idea was planted, which I went back to over the next few years as I persisted with administration work, until I considered working in the horticulture industry. It would combine many of the things I enjoyed in life, the outdoors, helping people, growing things, observing the seasons. So I commenced running a gardening and lawn mowing business...*

The concept of self-efficacy is important for adult learners generally, and will be discussed further in a later chapter of this book, in which we bring together the common threads that link our stories. Sufficient to say here that, according to Eraut (2004a), "if there is neither a challenge nor sufficient support to encourage a person to seek out or respond to a challenge, then confidence declines and with it the motivation to learn" (p. 269).

All three of the case stories in this chapter illustrate a willingness to meet new challenges in science and technology. The three people concerned, however, met these challenges differently. For example, at the latest point in his career Hugh deals with a wide range of incidents requiring informed analysis and synthesis:

> *I am currently part of a small team located in our London office dealing with all kinds of accidents and incidents on behalf of Insurers. Typically the losses we deal with cover a wide range of sectors within the aviation industry including drones, gliders, microlights, general aviation aircraft, airliners and even pre-launch satellites.*

He is required to bring experience and past knowledge of science and technology to each accident and integrate those elements into a coherent and practical explanation. Thus he is integrating formal learning with the implicit and tacit learning gained from long experience. According to Eraut (2004a, p. 256), using a scientific concept in a practical situation requires it to be "transformed or resituated in a form that [fits] the situation", and this is not "a process of logical reasoning that could easily be retraced but rather one of mulling over the situation until something seemed to fit". Further,

> in most situations several different areas of knowledge had to be combined, creating a problem-solving situation resolved by insight, which again could not be explained easily. Hence we have the paradox of professionals being able to refer to codified, scientific knowledge in clear explicit terms, yet using that knowledge in ways that are still largely tacit.
>
> *(p. 256)*

Ketan integrates his knowledge and experience theoretically to seek understanding about the issues arising with sections of the public who do not support what he is doing. This then has become a major focus of his work, supplemented by writing in social media:

> *I also soon realised that there were good scientific reasons for the prevalence of the "wind turbine syndrome" phenomenon. People who misperceived the risk of a technology would often express their fears through a narrative of helplessness and disenfranchisement. In countries where communities were deeply involved with wind projects, wind turbine syndrome was unheard of. This was the Saganist in me, telling me that although the rejection of science may be undesirable, it happens for a reason.*

Finally, Keith – although he says *"I would never have thought of me as, say, a doctor"* – is in fact the closest of the three to the way in which medical practitioners learn.

> *When I have a question in my current work I don't usually consult the web – I have a horticulturalist I speak to when I have a problem. There is a logical approach to solving problems. For example, I helped establish a native garden but there was poor growth and lack of nutrients – it turned out to be root rot from overwatering. How do you get to know that? I am not sure, but I go to my adviser and tell him what I have observed. You become a sponge in a rich environment.*

This kind of reactive learning has been well described by Slotnick (1999) and others with respect to the medical profession. Keith is, in effect, behaving as a diagnostician faced with a new plant problem. This corresponds to Slotnick's explanation of how a physician, when confronted with a specific problem, seeks answers in existing medical sources. "Generalizable knowledge is not the primary motivation" (p. 1107). If the problem is unanticipated, learning may be "done *ad hoc*". In this case, the physician will consult "immediately available resources such as knowledgeable colleagues and journals" (p. 1107). Finally, "the physician applies the learned solution to the problem and sees what happens" (p. 1109). This is precisely what Keith does with his plant "patients". Like many others, he is unsure how he came by his knowledge, thereby demonstrating the value of tacit and implicit learning for this kind of diagnosis.

The importance of mentors is clear in the cases of Hugh and Keith. Both acknowledge the role of mentors in their past and current employment. The value of mentorship has been investigated for many different professions, generally at the early stages of a career, but in Hugh and Keith's cases, mentors have been valuable over many years. In Ketan's career, it is role models who have played a significant part in his early fascination with science communication. The use of mentors and role models to encourage learning in science will be further discussed in the final chapters of this book.

References

Bandura, A. (1982). Self-efficacy, mechanism in human agency. *American Psychologist*, 37, 122–147.

Bandura, A., Barbarinelli, C., Caprara, G. V., & Pastorelli, C. (1996). Multi-faceted impact of self-efficacy beliefs on academic functioning. *Child Development*, 67, 1206–1222.

Eraut, M. (2000). Non-formal learning and tacit knowledge in professional work. *British Journal of Educational Psychology*, 70, 113–136.

Eraut, M. (2004a). Informal learning in the workplace. *Studies in Continuing Education*, 26, 247–273.

Eraut, M. (2004b). Transfer of knowledge between education and workplace settings. In H. Rainbird, A. Fuller, & H. Munro (Eds.), *Workplace learning in context* (pp. 201–221). London: Routledge.

Reber, A. S. (1993). *Implicit learning and tacit knowledge: An essay on the cognitive unconscious*. Oxford: Oxford University Press.

Slotnick, H. B. (1999). How doctors learn: Physician's self-directed learning episodes. *Academic Medicine*, 71, 28–34.

7

LEARNING THROUGH A DIVERSITY OF APPROACHES

The Case of the Moon Diary

The previous chapters have illustrated the different paths down which people pursue scientific knowledge. In all these case stories, the learning the authors achieved was a consequence of their own motivation and self-direction and was generally pursued in an informal manner. Many adults, however, choose to enrol in formal courses, whether these are delivered through an institution such as a university or technical college, or through informal learning opportunities such as the ubiquitous "university of the third age". Learning about science in these circumstances is, of necessity, learning about themes and topics that are generally decided by others. In the case of formal institutions, there will be a formal curriculum whose stated outcomes are required to be attained in order to pass the course.

How, then, can the passion and drive associated with intrinsic motivation be translated into a more formal environment? Given that learning is highly personal and, as we have seen, often highly selective, the message for those who wish to communicate matters of science to a general audience is that people will prefer to follow their own personal pathways. For those professionally engaged in communicating science, whether through an adult learning establishment, a scientific research organisation, a museum, or popular media, it is critical to understand that approaches to learning are essentially personal. This chapter describes an attempt to persuade prospective science communicators of the reality of this personal approach on the part of a possible audience, and the diversity of approaches to learning that it engenders. At the same time, it examines if there can be a conversion from extrinsic to intrinsic motivation within formal course requirements and constraints.

Over the past few decades, there has been a rise in tertiary programs in science communication. Students in these programs have a strong science background but are novice communicators who, for various reasons, have to reach audiences such as farmers, environmentalists, health professionals, and so on, including the general

DOI: 10.4324/9780429434303-7

public – anywhere, in fact, where scientific issues intersect with public life. In such tertiary science communication programs, the aims for teaching and learning are usually centred around individual growth and understanding on the part of the student, particularly with reference to understanding the various stakeholders with whom the novice science communicator might interact in the future. It is these novice communicators, and their need to understand audience diversity, who are the subject of this chapter.

The Influence of Learning Styles and Multiple Intelligences

A major influence on how adults set about learning scientific information will be their personal learning style. Much has been written about the concept of learning styles and the different ways in which people learn (Hawk & Shah, 2007). The frequently stated implication of this research is that teachers, particularly of adults, need to bear in mind this diversity. For everyone involved in education, the biggest personal hurdle to overcome is the implicit belief that everyone else approaches a task in the same way as ourselves – but a learning experience constructed along lines that we believe to be effective will not necessarily prove to be so for all learners.

> Rather, because the premise is that adult students learn in different ways, faculty in higher education would have a responsibility to expand their repertoire of learning activities to embrace as wide a field of adult student learning styles as possible in order to achieve more effective learning.
>
> *(Hawk & Shah, 2007, p. 2)*

Hawk and Shah made the claim that informed use of learning styles will result, for adult students, in increased satisfaction, performance, and learning, both during and beyond the course (p. 18). Various learning style instruments focus on different aspects of personal style. In general, many learning style inventories aim to identify preferred characteristics of engagement or experience with learning, for example, they may classify learners as active, reflective, theoretical, or pragmatic (Honey & Mumford, 1982). These are very useful ideas to help a teacher or facilitator devise various appropriate experiences and activities to underpin specific prescribed content. Such classifications of learners by their preferred learning experiences, for example, Kolb's Experiential Learning Model (Kolb, 1984) or Honey and Mumford's Learning Styles (Honey & Mumford, 1982) are not, however, the main focus of this chapter. Rather, different inventories which classify learning styles into categories such as Visual, Kinesthetic, and so on are more useful for this discussion (for example, aspects of the Felder and Silverman [1988] Model). In particular, the Theory of Multiple Intelligences (Gardner, 1993), while different from learning style classifications, is relevant in cases where people learn within a highly personal, independent framework.

According to this theory, there are at least eight basic intelligences which, given a free choice, will tend to dominate learning. In summary, the main eight intelligences of interest in this chapter are:

1. Linguistic intelligence. Learning through reading, writing, narrative.
2. Logical-mathematical intelligence. This incorporates logical (causal) reasoning and the ability to understand abstract concepts and numbers.
3. Spatial intelligence. Visualising, making mental pictures.
4. Musical intelligence. Sensitivity to rhythm.
5. Bodily-kinesthetic intelligence. Using the body to solve problems: hands on experiences, animations, simulations. A concrete approach.
6. Inter-personal intelligence – the ability to understand and work with others. Effective communication skills.
7. Intra-personal intelligence – the ability to form an accurate model of oneself and use it to operate effectively. Self knowledge.
8. Naturalistic intelligence – learning through contextual applications of the natural world.

These intelligences, according to Gardner, are distinct. Learning styles, on the one hand, illustrate how an individual approaches a particular task. Gardner's intelligences are not "learning styles"; rather, they reflect aspects of intellect. For any individual, one type of intelligence will have more appeal than another depending on the degree to which that particular intelligence is developed. For the most effective learning to occur, relevant experiences embodying that intelligence must be provided for. This does not imply, however, that learning cannot occur through other intelligences or that a person may not have more than one preferred mode. Gardner himself resisted the idea that the theory should be used to lock learners into a specific style; rather, he said that the theory should be empowering and assist the learning process. Strauss (2013) cited Gardner's own words, as follows:

> As an educator, I draw three primary lessons for educators:
>
> 1. *Individualise* your teaching as much as possible. Instead of "one size fits all", learn as much as you can about each student and teach each person in ways that they find comfortable and learn effectively …
> 2. *Pluralise* your teaching. Teach important materials in several ways, not just one … In this way you can reach students who learn in different ways. Also, by presenting materials in various ways, you convey what it means to understand something well …
> 3. *Drop the term "styles"*. It will confuse others and it won't help either you or your students.
>
> *(Strauss, 2013, original emphasis)*

Despite criticisms that this approach has not been proven (see, for example, Waterhouse, 2006, who states that there are no empirical validating studies) the idea of Multiple Intelligences has resonated with the education community. Critics claim that it is, to some degree, a self-fulfilling prophecy when it is successfully applied in the classroom. Nevertheless, in this chapter it is referred to as a useful tool to analyse the educational experience of the "Moon Diary", an assignment given to a group of students in a Master of Science Communication course. The aim of this assignment was, as stated above, to foster understanding of the diverse perspectives of the various stakeholders with whom the novice science communicator might interact in the future.

Free Pathways and Motivation

What does it mean to give adult learners a free choice within a given topic to follow their own pathway to learning? According to Deci, Vallerand, Pelletier, and Ryan (1991, p. 342), "promoting greater self-determination, that is, a greater sense of choice, more self-initiation of behavior, and greater personal responsibility, is an important developmental goal". Self-determination is important in fostering creativity and "cognitive flexibility".

Various authors have described a person's actions and behaviour in terms of motivation. Intrinsic and extrinsic motivation have been recognised and explored for many years and they were described in detail in Chapter 2. The distinction has implications for adult learning. Self-determination theory is not only concerned with processes of directed behaviour, but explores reasons why an outcome might be seen as desirable. It may be, as in the case described in this chapter, that successful completion of a task will gain good marks in a formal course. Even if the task is thus extrinsically imposed, a high degree of self-determination is nevertheless likely to enhance the learning and increase motivation. This is an example of "integrated regulation"; "the regulatory process is fully integrated with the individual's coherent sense of self" (Deci et al., 1991, p. 330). When a person believes that she or he has choice in what is to be learned, then self-determination is the governing principle. If there is an externally imposed goal, as is usually the case in formal adult learning programs, then the degree to which the student identifies with the goal as intrinsically motivating is likely to affect the outcome.

Designing adult learning programs in science with these ideas in mind presents problems. Deci et al. (1991) suggested that such a program includes "offering choice, minimizing controls, acknowledging feelings, and making available information that is needed for decision making and for performing the target task" (p. 343). This is not easy.

As discussed earlier in this book, the "Andragogy in Practice Model" (Knowles, Holton, & Swanson, 2015, p. 6) incorporates three elements: considerations of goals and purposes for learning; individual and situation differences; and core adult learning principles. Over the past few decades, the importance of individual differences among learners has become more obvious to researchers in the field. "The key point is clear – individuals vary in their approaches, strategies and preferences during learning activities" (Knowles et al., p. 86). The "Andragogy in Practice Model" developed by Knowles and colleagues in 1988 and quoted again in the 2015 edition (p. 6) has the following steps:

- Learner's need to know: they must understand why they need to learn something.
- The self-concept of the learner must lean towards self-direction.
- The learner's prior experience should provide a resource for learning.
- There must be a readiness and willingness to learn.
- The learning is problem centred and contextual.
- Motivation is intrinsic.

This last point presumably can include integrated regulation. The first condition in the list above can include the case where the need to know is prompted by an external stimulus such as a valued outcome. In summary, therefore, adult learners will do best when they have some control, not only over what is learned but how it is learned.

The Moon Diary Assignment

In science communication theory, there has been considerable attention paid to the shift in approach from the deficit model of "educating" the public to a model of dialogue in which the public is an equal partner in the exchange (also see Chapter 1). If a communicator of science wishes to engage adult members of the public, then those adults must be given as free a hand as possible to follow a pathway suited to their own interests and needs. This concept can often be difficult to grasp, particularly if the communicator comes from a strong scientific background in which "facts" are the obvious means of communication. Many such communicators embrace a deficit model and assume that everyone can understand and learn through a transmission model of information. Changing this approach is a key element in science communication studies.

Tertiary science communication courses address this problem through reading and instruction. It often proves problematic, however, to move students with a strong education in science and scientific process to a position of understanding the diverse approaches they must adopt when communicating with an adult "lay" group. Such a group might be the "general public" (a catch-all term, because we know how different our publics really are) or specific stakeholders.

In one science communication program at the Masters level, this aspect of science communication and engagement has been addressed through a device known as the "Moon Diary". Adult students in this program are all science graduates and prospective science communicators with a range of employment aspirations. Some are already in positions within scientific institutions or government that require them to communicate science to stakeholders or the general public. Some of their stakeholders may be politicians, often Ministers of Government. They may be farmers or environmental workers. They may be people seeking information about health or other personal matters. It is therefore critical for these students that they understand how to communicate their science with a deep understanding of the multiple contexts within which their prospective "audience" may interpret the information. For those who aspire to careers in the media, they too need to know how to vary the context of science information.

It is easy to tell the students about this diversity but, as stated above, this does not usually make the point strongly enough. The Moon Diary assignment provides a graphic and often surprising demonstration. Even though the students are all science graduates with honours degrees in their discipline, the variety of approaches to the task, which emerges from the group, crosses disciplines and cultures. This reinforces the message that their "audiences" will also take and use information in different ways.

The Moon Diary assignment is discussed here because it embodies many of the principles outlined in the first part of this chapter. It promotes greater self-determination, that is, a greater sense of choice, more self-initiation of behaviour, and greater personal responsibility than conventional learning. The learners begin to understand why they need to learn about adults learning science, while the assignment itself inclines to self-direction and relies strongly on prior experience. The learning is essentially contextual. The assignment is, therefore, an example of adult learning in which there is a choice of roads to follow.

The task is set within a unit of study called "Communicating science with the public". The task forms part of the formal assessment for the unit, accompanying other assignments such as a group research project, a portfolio presentation, and so on. At the outset, it is explained to students that the purpose of the task is to understand the problems and pathways facing the general public when they become interested in a topic but have little background in science. The topic in question is the Moon. This was chosen because it is relevant to everyone, is observable, and is a subject about which there is a large quantity of information available. Somewhat surprisingly, it is also a subject about which many people know very little. The task is described very simply:

> This semester, keep a diary about the Moon. You may include any material you find interesting: there are no guidelines. The only rules are that you must present your diary to the class at the end of the semester, and that we will not discuss the project in class at any time. If you wish to discuss it between yourselves, that's fine – the public have that option too! A major purpose of the Moon Diary is to reflect on your own learning and you are expected to include such a reflection when you submit your diary at the end.

There are generally two kinds of immediate response. Some students see the project as immediately liberating ("do you mean I can do whatever I like?"). Others are worried and confused. There is a third group who do not comment immediately – these are the "wait and see" students who need time to reflect on the task and assess what they need to know.

Except for answering a few questions of clarification, there is no discussion about this project until the presentation day, on which each student is asked to describe their approach to the project and talk about the product. This assignment has been given for over fifteen years and the outcome is always surprising. Despite their common background in science, each student takes this project along a unique path, and the examples which follow have been drawn from the diaries submitted for assessment over this long period. The overall cohort was approximately 200 students, of whom these are a representative sample.

Initial Responses: From Confusion to Elation

Many students react initially with confusion and, in some cases, resentment. These are, usually, high achieving students who have always done well with firm assessment guidelines and are unhappy at being set adrift in this way. They question how the task would be assessed and are uneasy at the response that – *really* – anything would be acceptable providing they produce the reflection on their learning.

> *The entire concept of the moon diary confuses me. While I have written diary and journal entries before I have never done so without set guidelines. My education up until this point has been about following the rules and structure set for me by my lecturers. (S1)*
>
> *I am here because I have to be. A strange assignment with no rubric, criteria, or guidelines dictates that I must "keep a moon diary" whatever that means. No further restrictions or instructions. Some would find this breadth of scope refreshing; I, however, find it constrictive. The very idea of having no guide means that I am terrified to write anything lest I "do it wrong". I certainly don't like not being good at things… In the last few minutes, having started this blog, I have also come to the realisation that I can use this task to return to something I dearly love and rarely do: writing. Over the years I have certainly come to think of myself as someone who is not creative (oh, how much power these paradigms hold!) but recently this self-conception has been holding me back somewhat. So: the moon diary can serve a purpose, to allow me free and prolific pontification. Who knows, I may enjoy it so much that it becomes a habit. (S2)*

Later, in the final reflection, S2 was to write:

> *Have I been academically successful because the structured environment fits my natural learning style, or have I been so conditioned over the years to work in this environment that I now cannot cope with anything else? This is something I never thought about*

> *until I was confronted with the Moon Diary task, which made me feel so uncomfortable. I hope that after completing this task, I will be less afraid the next time an unstructured task is presented to me, and I will be able to give it my best effort, as I have done with the Moon Diary. (S2)*

In a similar vein, S3 found that the task was initially confronting, but later enlightening:

> *When given the task of creating a Moon Diary, I was initially nervous and concerned about the direction it would take. My schooling until this point has been extremely structured and the assessment tasks I have been given were well defined – this was the opposite. After thinking about the task, I realised that this was the point. (S3)*

It was important to acknowledge to these students that their feelings of insecurity were entirely legitimate, as mentioned by Deci et al. (1991, p. 343). The level of concern expressed by these high-achieving students was stressful for them and, in some cases, resulted in angry outbursts. It is notable that, for many science students, the learning pathway has been so highly structured that it has never given them options or alternatives and they are thus ill-prepared for free learning and for understanding what this means. They are frightened by the concept. Many students, however, who recognised this lack of structure as a personal problem decided to self-impose guidelines of some sort on the task. In general, this settled their concerns and gave them a basis from which to construct the diary.

> *I'm not used to having so much freedom with a task. In the past, all my assignments, essays and assessments were very rigid and defined, in both their guidelines and structure. I was always required to follow the criteria and rubric, and simply answer the questions. (S4)*
>
> *I would say the moon diary task was particularly challenging for me. It was a foreign experience to be given complete freedom in an assessment task and frankly quite daunting... However in thinking of the task as a learning experience, it forced me to trust in myself and my own convictions in putting together a piece of work that I felt was worthy. I also had to choose my own criteria for assessing my success and to decide how much and to what level I would research the Moon. In my research I tried to follow my interests so as to engage with the task of learning about my own way of learning. (S5)*
>
> *I really struggled at first to start the moon diary. It took me 3–4 different attempts. Without a structure or guideline I really struggled on where to start. I couldn't go and then change my mind half way through. So what I ended up with were basically the bits I like the best out of the 4 or 5 attempts. I created my own structure. Once I did this, I rather enjoyed the task. (S6)*
>
> *I am freaked out about the lack of guidelines and structure of this assignment. I have no idea what [they] are really looking for. I can begin by saying one thing. My complete lack of understanding of this assignment has already taught me something about the way I learn. I need clear logical instructions, and perhaps I'm not very good at being responsible for my own learning.... Entry two: I think I get it now. It has been*

about two weeks since I've written. After much frantic discussion and complaining with my fellow [students], I have come to the conclusion that everybody has a different interpretation of what The Moon Diary is. Those who have decided what they will be doing are all doing something different. Some people even think that it's not even really supposed to be about the moon. But talking it through with my peers has helped me to come to the conclusion of what I think it is. Now that I know I'm a person that needs clear, logical instructions, I'm going to set out my own goals and guidelines for the diary. (S7)

For others, the freedom to follow their own pathway was immediately liberating:

Yesterday in class we were given the task of writing a Moon Diary. Of course, my first question was, and still is, what is a moon diary? However, it appears that the answer to this question is elusive. There are no answers to be had from the course guidelines, or from my peers. Nobody knows what a moon diary is. People have theories and ideas – some of them quite predictable, others quite creative. Herein lies the beauty of this assessment piece – we can make the moon diary whatever we want it to be. It is a space for learning, personal development and reflection. I can make it whatever I want it to be. (S8)

The only thing I know for sure about this task is that it is designed so that I can learn more about the way I learn. BUT THIS EXCITES ME!! (S9)

At first, the lack of structure really overwhelmed me, but even that helped me discover how I like to learn. When I passed the overwhelmed stage I realised that the beauty of the assignment was that I could choose to play to my own strengths. (S10)

The difference between this Moon Diary and the "normal" assignment set for university students in common science subjects is stark. I believe this to be the reason for the extraordinary and excitable reaction from students in the class on receipt of the assignment brief. The discomfort displayed by numerous individuals provided a valuable insight into the institutionalised development of the need for instruction from course coordinators and the associated comfort derived from order and direction. The worry that individuals may fail due to the perceived risk of misinterpreting the brief was immediately evident. I enjoyed it. (S11)

Choosing the Theme

Having decided to embark on the diary, students chose a variety of approaches.

1 The Science of the Moon

A number of students decided to look at facts about the Moon. The most "mathematical" diary was probably the student who decided to take measurements and derive equations governing its behaviour from first principles.

Observations, in many forms, were recorded: one student engaged family members in different parts of the country to record what they saw. Tides, eclipses, and misconceptions all formed frameworks for looking at the science. Students often acknowledged their lack of knowledge:

> *After my research, I now realise that I previously held some misconceptions, and also had considerable knowledge gaps, about the moon and its phases. I believe that I am someone who is well informed about a range of topics. I have a sound scientific background, and good knowledge of general science topics. And yet I had no idea about how this everyday phenomenon – one which I have observed hundreds of times in my own life – actually works. (S12)*

It was notable that many students who decided to keep a fairly literal diary of the Moon had little understanding of its phases.

> *I learnt that the relationship I have with the moon is built on assumptions! At the start of the task I assumed that the moon would be there every night. Old trusty moon, easy to spot, looking back down at me… If I had thought about it at the time, I did know about the phases of the moon, so I could have worked out it wouldn't have always been there, but something in the back of my head was telling me I'd always see it. Although I had learnt that the moon would come and go in the sky, my belief that it would always be there still overruled the logic that I knew to be apparent. I combined these ideas in my head to think that I was going to see the moon, every night, in the same spot in the sky. (S13)*

One student, who shared a house with other young people, went home the first evening and found to his consternation that no Moon was visible that night – or the next several nights. He recounted, on presentation day, that he had said to his housemates, "It's a trick, right? There is no Moon here?" to be met with scorn. The outcome was that everyone decided to help – his house friends, and his Facebook friends, all became engrossed in the task and would draw pictures for his diary.

The facts of the Moon extended to investigations of the Space Race and conspiracy theories, and to taking online quizzes to widen Moon knowledge. Aspects of human behaviour, including menstrual cycles, were also researched.

> *I want to find out if the moon's cycles can affect human behaviour. Or is it just a myth? I will use this diary to journal my progress and reflect upon my learning at the end. (S14)*
>
> *Reflecting back on my moon diary, I can easily say that I enjoy learning facts. I noticed that I had a preference for videos that were the most novel and entertaining compared to others that I could choose from. I wrote in my diary entries that I was more engaged and learnt more from these videos than those that were not novel. (S15)*
>
> *To produce my Moon Diary, I monitored the activity of an online public forum … for a month. Over the course of the month, I pulled out questions that the public asked regarding*

the moon... For the Moon Diary I looked at the answers to the questions that had been "upvoted" the most – the answers with which most people agreed. I would then dissect these answers into separate scientific points. This was to enable me to state whether I agreed or disagreed with what was being said as easily as possible; I wanted to prevent myself from not taking a stance on any given issue. Once I had done this, I was able to look up the information in journal articles and textbooks to either support or debunk what the answerers and I had said. (S16)

As a member of the public, I personally have two main methods of finding out new scientific information; ask the Internet and ask my friends. I decided to investigate the latter option for this Moon Diary and interviewed a small group of my friends about the moon. I recorded our discussion and later analysed how we'd communicated. The group was made up of mathematicians and engineers. The discussion wandered across many different topics related to the moon, including; gardening, mental state, weather, tides, gravity, the origin of the moon, films, missions to Mars, claustrophobia, sleeping, experiments in space, and economic viability of space travel. In general, the group tended to share interesting facts or stories they'd heard, pose them to the group and try to come to agreement on the validity of the story based on likelihood and common sense. Each member usually contributed some component to the discussion to create a shared understanding and consensus for the group. If someone shared a story that was new or of particular interest, the group members would all begin talking at once as they assimilated the news. Usually the group took care to limit the level of tension and distress to any individual member of the group, often qualifying their opinions or ideas with an "I don't know" to be inclusive, posing statements as questions and changing the topic when discussion got tense. (S17)

This last student later stated in her reflection (see later) that she really only understood the point of the assignment at the very end. Nevertheless, she turned the task into something that she enjoyed, by focusing on the communication that was occurring, rather than the facts. A similar outcome occurred for student S18, who turned the task into a communicative and social exercise:

It also meant that I could achieve something productive while doing something I enjoy – that is, talking with friends and acquaintances. The running theme for my moon diary was the theme of "moon facts" – whenever there was a lull in the conversation, or I needed a conversation starter, I would ask people for moon facts for my moon diary. In some cases this helped me to start friendships with new people, such as when I asked others in the choir I had just joined whether they knew any moon facts. I would then jot these down using my phone – another new skill for me, as I only recently bought a smartphone and have been learning how to use different functions such as taking notes .... Through these discussions about the moon with lots of different people, I feel I have picked up many interesting facts about the moon, but much more importantly, I have learned how to tap into the expertise of other people as a resource to gather relevant new information. In particular I learned that people are usually keen to share what they know, especially if they think it will be helpful. This tied in well with several of my evidences of learning, which showed me how community-based science programs can be highly successful for improving public awareness and perceptions of science. (S18)

2 The Visual Moon

Within the group who looked at the Moon fairly literally, several opted for visual methods of recording their observations. Either they drew what they saw, or took photographs.

My original aim for my moon diary was to take photos of the moon with the idea that I'd slowly progress on my ability to take photos.... I took little annotations initially to jog my own memory of the events of that day and why the photos either turned out good or bad. (S19)

I needed to learn something new about the Moon for this to be a rewarding and fulfilling task. I have always wanted to know more about the space race and how the technology was developed to allow space travel. I am also a visual learner, I learn best when there are pictures or diagrams to complement pieces of information. I find researching information to be enjoyable, but I tend to learn short, interesting facts rather than long bodies of text. So I combined both of these methods and decided to make a scrapbook of information about the space race. (S3)

Student S3 who had described a high level of uncertainty at the start (above) thus found satisfaction in the decision to go with her perceived strengths.

Moon videos were popular, and many students presented their diaries with a wealth of photographs of the Moon at different times and phases. Blue moons, too were explained. The famous *Earthrise* image featured largely in many diaries.

I never thought of myself as a visual learner, like I said before I rote learnt a lot of my way through school, however I really enjoyed looking at pictures of moons and this is often how I chose which moons to do in the diary, also if it had a quirk or a little fact I found interesting. (S20)

3 The Moon in Music

Some students (a minority) represented their Moon Diary in musical form. One student composed a piece for voice and harp, thinking about the Moon's effect on mood. Another decided to map music onto the phases.

I like being creative so when the moon diary was discussed I knew I wanted to represent the moon in an artistic piece. Music is a passion of mine so I decided I would reflect the phases of the moon in a song. I originally thought of composing a new piece with the eight phases signified by new movements but without access to a piano I decided that would be too difficult. So instead I decided to sort through all the sheet music I had and pull out music that reflected the moon in some way. This process relied on creative license as some of them are related to Heaven, Angels, Eclipse, Night, Dreams, Planets, Stars, Twilight... Photos were then found to show the phases of the moon as well as poetry about each phase. These were combined into a movie which can be found on YouTube 21. (S21)

For the moon diary I have taken the approach of doing series of sketches of the things I associate with the moon and writing down all the random things that spring to mind, and either tracing them back to their source or following them up on them until I have a satisfactory answer. I know now that I tend to associate music with everything, this is in part because I grew up with music much of it live around the house all the time so that almost any memory from my childhood has music in it, I tend to study using music as over the years I've noticed that my factual recall is way better if I involve as many senses as possible and because I'm a lyric listener when it comes to songs. I tend to listen to the words. (S22)

4 The Moon in Film and Fiction

A large group of students wrote stories about the Moon or looked at its role in film. Most of the stories were for children, often illustrated.

I wanted to present an activity book for very young children, something in which they could look at the moon each night, learn the different phases and record data, the most basic of experiments. This idea initially was because I have a two-year-old who loves to find the moon each night and describe its shape. However, as I started to read through the information on the moon, I soon realised no two year old would understand the phases of the moon; my son calls a crescent moon, a banana and a full moon, a circle. There are only two phases according to him. The moon diary soon became targeted at older children, mid primary school age, comprising a few pages of information that would interest them... I am definitely a visual, hands on style learner and find data, graphs and loads of writing off putting. I kept the information very short and as simple as possible... I did ask a few primary school children to read the booklet, with some great feedback from them. More jokes, more animals, more machines ... so I did add two more jokes and some more facts. (S23)

On reflecting through the diary, I noticed I had no pictures in it. As I went through and added some in, I realised I had not added them in as I went, because I was not as visual a learner as I had once thought. I was clearly a very linguistic learner, as I engaged more with the narrative and story aspects, than the associated pictures, for example. (S24)

Others chose to address the diary directly to the Moon "herself", pondering what they saw and felt about the Moon at the time. Some invoked poetry to convey their feelings, or described poetic thoughts:

Thus one day, I suddenly came up with the idea to write something about "spending my honeymoon on the moon". It actually was one of my dreams when I was a senior school student. I knew it was impossible to spend my honeymoon with my future partner on the moon, it costs too much and there are some other requirements, for example the physical test. But at least I can write something about it. (S25)

5 Misconceptions About the Moon

The group had discussed the idea of misconceptions during the regular class, and several decided to investigate this aspect. The form of the investigation varied: one student made a video interviewing people about their ideas while others constructed a survey.

I set out to ask people about the phases of the moon, with a focus on determining what it was that they didn't know, and why. What I ended up with, however, was a far more interesting problem than simply general lack of knowledge. As has come up in class time and time again, it was people's attitudes to lunar science that were the most worrying. Those I was interviewing – 28 people, in alignment with the number of days in the lunar cycle – were consistently nervous and uncomfortable about answering a "science question". So this became the focus of my moon diary. I began to document the shift in attitude, in posture, in facial expression, and the outward distaste people showed when faced with the task of answering a science question. Specifying before every interview that there were no "expected" or "incorrect" answers, I asked what people's immediate association were when they thought about the moon. Common responses were largely pop culture related – the man in the moon, "I used to think it was made of cheese" and so on. While talking about personal experiences, all interviewees were comfortable, amiably chatting about a wide range of association, both scientific and non. All of this, however, was reversed once science was brought into the picture. (S26)

6 The Moon and Culture

The question of cultural views of the Moon was popular with many students, particularly those of non-Caucasian background. Many researched different cultural images and stories and often related these to personal backgrounds. One student, who was interested in history, connected the "rabbit in the Moon" to her childhood. The image of a rabbit or a hare is common to many countries, including South Africa, Japan, China, Korea, Singapore and even First Nations peoples in North America. The celebration of "Moon cakes" at the mid-Autumn Moon festival, which also featured in some accounts, is common to Chinese and Japanese cultures across the world.

Being in a small village in the middle of summer, I could hear all of the neighbours talking excitedly as they sat in their gardens or patios, eating traditional moon rice cakes, complaining of the heat... There was a lot of cloud cover that night, so we were not optimistic as we waited for the time when the moon would be at its highest point in the sky. Miraculously, as the time approached, the clouds cleared, and I saw the brightest and whitest moon I had ever seen. And for the first time, I could see the image of the rabbit making rice cakes on the moon. (S27)

7 Random Thoughts

Given the open-ended nature of the assignment, there was a significant group who elected to let the diary simply "flow". Often, however, the thoughts were not completely random, but linked in some way either to a theme or to the course itself:

My plan for this moon diary is to experiment with the idea that after becoming aware of something, you hear about it everywhere. Therefore the experiment I am proposing with my Moon Diary is to not actively research information about the moon, but to see whether I will gain knowledge through this phenomenon, but also reflect on what I already know about the moon. To further complete this jigsaw of knowledge, I will have conversations with the people around me, my trusted sources of information, to also fill in these knowledge gaps. I am aiming to complete the jigsaw without needing to resort to the Internet to find an answer. Obviously I don't foresee that all aspects of the moon will be covered by chance encounters over the next few months, instead that I will gain new knowledge about a range of different aspects of the moon. (S28)

After investigating these I have decided that my moon diary will be a combination of information about the moon and its phases (drawing from the lunar calendar), and also a reflection on how my learning in this subject (and science communication), is progressing in general. I think it will be interesting to try to make links between the moon and my own life/learning. (S29)

My approach for this diary will be to start writing and see where my thoughts take me. I will endeavour to present interesting facts, observations and ideas based on my experiences with the moon and my feelings towards this comforting component of our solar system and its reliably constant and predictable changes. Most importantly I will be attempting to deliver learnings on the moon that I think have or could have been presented to the world based on the concepts learnt throughout this course. (S30)

Reflections

Everyone was able to reflect on what the exercise told them about their own learning. The open-ended nature of the task, although initially confronting, was later understood to have allowed students to follow their own directions and to understand themselves more in the process:

> *Although I was initially nervous about the Moon Diary, I found the task to be very rewarding .... The freedom enabled me to express myself and confirm to myself my most effective learning style (S3).*
>
> *Let's start at the beginning, with my struggle to understand what was required of me to do this assignment. This was my first entry of the diary. I complained profusely about what this task meant. What did the markers want? I thought it was outrageous that there were essentially no guidelines. That is not how I've been conditioned in my scientific education to go about things. There needs to be an aim in everything you do. This is how I thought and this is why I struggled to know what I should do .... Upon reading this first entry back in February (and even at the time), I had already come to an important conclusion about myself. I needed direction with my learning. I needed structure. I needed a goal because I didn't like the open-endedness of this task. It took weeks of discussions with my classmates to actually help me to come to sort of understanding about what was required. This was also documented in my first entry as it highlighted another important fact, which is that I find it essential to learn socially, so I can draw on other's ideas to create my own understanding (S7).*

For others, the task enabled a greater understanding of how they preferred to learn and, perhaps, that this might need modification:

> *Throughout this moon diary exploration, I have become painstakingly aware of how much I enjoy simply "jumping in" without much consideration of how things are going to turn out. From the outset, I had in my mind a clear idea of what I wanted people to say, what I expected them to say, and how I expected them to react. I enjoyed the minimal constraints and boundaries put forward for this task, and thought that this directly related to the freedom I would experience in its undertaking. This was, of course, not the case; while the assignment may have been open, my mind was not. My biases towards, and disregard for, my direct audience was an extremely inhibiting factor. I have learnt in the undertaking of this task that, despite thinking I know my audience relatively well, I still have a great way to go in terms of understanding the "public". (S31)*
>
> *So, what has this taught me about my own learning? Well firstly... I would say that I am first and foremost a linguistic learner. That is, I like to deal with words and language and learn best through reading and taking notes. This was clear from the way I chose to approach the task .... even though I'd refer to diagrams sometimes to aid my learning, I found I couldn't really understand them that well without some sort of written explanation. (S32)*
>
> *I have also discovered that I enjoy learning in different ways... I like learning (and communicating) through both text and visuals (photos and pictures). I am less inclined to look at graphs, complex diagrams, and mathematical formulas. I also found that I have equal passion for science and the arts, and that these two fields do not have to exist in isolation, but can rather work together to be powerful science communication tools. This is something that I can use to my advantage as I continue on my science communication journey. (S8)*

> *It was not as hard or unpleasant a task as I suspected it would be .... It seems the Moon Diary, like the Moon itself, has had a broader and more significant impact than I first suspected. (S33)*

The overall purpose of the exercise had clarified by the end of the assignment:

> *The real surprise and the real learning experience came through the discovery that I have been so institutionalised that the very idea of completing a task without step-by-step guidance terrifies me half to death. How much of this is learned and how much is innate? (S22)*
>
> *What I have discovered about myself is that although I thought I liked structure and guidelines, I actually found it a lot more rewarding having a bit of flexibility. It has also taught me that I need to embrace my creative and imaginative side a bit more in the future. (S34)*

From the above extracts it will be clear that the students in this course took the task in many different directions and, as they all eventually realised, this was the point of the exercise. Not only did the students employ a range of strategies which might loosely be classified as "linguistic", "logical mathematical", and so on, but many also involved others in an inter-personal context. Group learning suited some, while others were introverted and were internally focused on the task.

The message of the Moon Diary experiment was clear to all the students at the end of the exercise. It was not necessarily clear to the students at the start, however, that this variety of approaches would be the outcome and, for some, the message really only became obvious on the day of the final presentations:

> *I'm just not enormously interested in the moon. I mean, its ok – it looks nice and all, but it doesn't really rivet me. Plus everyone had to do [a diary] so how boring must it be to have to hear about a thousand different facts on the moon from my classmates? On the day of the presentation of our moon diaries to the class, I went second and got mine out of the way... now I got to sit back and listen for the rest of the afternoon. Then I finally understood this assignment. Everyone had built something completely different. Due to the broad nature of the assignment, everyone found their own way of looking at the question and discussed their journeys in learning. It was fascinating to see the diversity in projects. I finally understood that what I thought was a fairly dull task was actually a brilliant demonstration of the differences in how we each interpret a task and pursue learning in completely different ways. As a Science Communicator, this is an immensely valuable lesson for shaping out messages and discussion to suit the varieties in how people learn. It was also a good reminder to reserve judgment on things as I may not have understood the full lesson they have to teach! (S17)*

Finally, the extended thoughts of one student underline and support the rationale for approaching adult learning in the manner recommended by Knowles et al. (2015) and Deci et al. (1991):

> *While the moon diary has been part of the coursework for Communicating Science with the Public, I found considering the assignment itself to have furthered my understanding on science and learning. The moon diary is an open-ended assignment with minimal direction. There is no "right", and no "wrong". This contrasts strongly with both the prevailing teaching strategies in formal learning environments and the popular view of science. In the formal schooling system, progress is marked by how much knowledge students can obtain, or get "right". This is seen in tests, exams, and even assignments where rubrics are tightly aligned to expected outcomes. In contrast, the moon diary anticipates that the learner is capable of constructing their own knowledge in the framework of their own learning and experiences. It does seem rather a "baptism of fire" – throwing students into an ambiguous assignment when all we have known is the relative safety of "right" and "wrong". Aside from the difficulty of knowing where to start, the moon diary begs the question – how do I know what I have achieved? Perhaps we have all been looking at the world through the wrong lens; there is no "correct". The purpose of learning is therefore not to know the answer, but to gain understanding and make connections with new information. This harks back to the complexity of constructivism; if each person constructs their own unique knowledge, is there ever a time when they are "wrong"? I don't know the answer, but it seems to lead to this: the achievement of doing something is not to be "right", or "best", but to gain something, however small, to add to your own understanding of the world. As for myself, perhaps this can temper my tendency towards perfectionism; "Have I learned something from this? Yes. Then it is enough ...." (S35)*

Thus the challenge for educators of adults in the informal or non-formal learning arenas – and, perhaps, the formal sphere also – is to provide for these elements of knowledge construction while still attaining the desired curriculum outcomes. We shall return to this dilemma in the final chapter of this book.

References

Deci, E., Vallerand, R., Pelletier, L., & Ryan, R. (1991). Motivation and education: The self-determination perspective. *Educational Psychologist*, 26(3), 325–346.

Felder, R. M., & Silverman, L. K. (1988). Learning styles and teaching styles in engineering education. *Engineering Education*, 78(7), 674–681.

Gardner, H. (1993). *Multiple intelligences: The theory in practice*. New York: Basic Books.

Hawk, T. F., & Shah, A. J. (2007). Using learning style instruments to enhance student learning. *Decision Sciences Journal of Innovative Education*, 5, 1–19.

Honey, P., & Mumford, A. (1982). *Manual of learning styles*. Maidenhead, Berkshire, UK: Peter Honey.

Knowles, M. S., Holton, E. F., III, & Swanson, R. A. (2015). *The adult learner* (8th ed.). New York: Routledge.

Kolb, D. (1984). *Experiential learning: Experience as the source of learning and development.* Englewood Cliffs, NJ: Prentice-Hall.

Strauss, V. (2013, October 16). Howard Gardner: "Multiple Intelligences" are not learning styles. *The Washington Post.* Retrieved from https://www.washingtonpost.com/news/answer-sheet/wp/2013/10/16/howard-gardner-multiple-intelligences-are-not-learning-styles/

Waterhouse, L. (2006). Multiple intelligences, the Mozart Effect and emotional intelligence: A critical review. *Educational Psychologist*, 41(4), 207–225.

8

RESOURCES FOR SELF-DIRECTED LEARNING

Four of the preceding chapters have presented very different learning stories for more than a dozen adults. Their journeys were sometimes short with a definite endpoint, such as Ana's pregnancy, although once it ended she began a new journey in learning to care for her baby daughter. Others are never-ending, such as Michael's continuing quest to create new and "unmake-able" pieces of jewellery. How these authors approached their learning task varied too, and this is not surprising, given the diversity of learning approaches people use in trying to make sense of what they need to know. As revealed in the previous chapter about the Moon Diary assignment, people attempting to come to grips with the task each had his or her own way of finding and interpreting resources required for completion. People accessed a range of resources with a variety of entry points (see Gardner, 1993, pp. 203–204) at various stages of their journeys. But all our authors had something in common: they wanted to learn and they persevered in seeking and using the resources they needed to achieve that learning. In this chapter, we begin a review of these resources, exploring how they might best be used and examining the features that facilitated learning for the diversity of self-directed learners. In the next chapter we focus specifically on learning from the Internet and new media.

How Self-Directed Learners Use Resources

We begin by summarising the different resources used by our case story authors to work towards achieving their goals. To make it manageable, the resources were grouped into like categories and Table 8.1 presents this summary of eleven kinds of resources, who used them, and whether or not they were found useful. Our authors' needs were different, they progressed at different rates, none found all resources equally useful, and one resource was rarely enough. For example,

DOI: 10.4324/9780429434303-8

TABLE 8.1 Resources used by case story authors

Resource	*Used by*	*When useful*	*When not so useful*
Books and other print media	Ana, Mary, Richard, Michael, Tina, Paulette, Paul, Liz, Tiki, Kristen, Warren, Keith, Ketan Hugh	Seeking specific information and find right book As everyday reference for identification, etc. Sources of contemporary information Easy access through libraries Can provide background overview	Too technical for user, or uses too much jargon. Too simplistic or superficial for user's needs
Television news and documentaries	Ana, Tina, Paul, Warren, Ketan	Provide specific information that is evidence-based Stimulate new interest or direction	Purveys opinion not information
Internet information	Ana, Penny, Tina, Paul, Liz, Tiki, Kristen, Warren, Ketan	Health and science websites selected on reputation Good for quick updates Good when used selectively and with cross-checking of sources Good for specific information	Technical terms often defined using other technical terms not properly understood. Information is opinion, not evidence-based
Social media	Tiki, Kristen, Ketan	Good when carefully vetted Contributing can be therapeutic	Purveys opinion rather than validated information
Experts in the field	Ana, Penny, Mary, Paulette, Kristen, Warren, Keith	Willingness to share knowledge Able to give answers to specific questions Sorting out what is important and what is not for specific issues	Problems understanding terminology Employees hampered in giving information by loyalty to employer
Taking courses	Ana, Michael, Tina, Liz, Tiki, Hugh, Warren	Focused on users' needs Provide relevant new information Covers essential material for expertise Access to equipment otherwise unavailable	Too basic for needs Too much of content not relevant
Teachers, mentors, role models, other educators	Ana, Mary, Michael, Liz, Kristen, Keith, Hugh	Specific, targeted information relating to knowledge and skills Modelling needed skills Point to further specific resources	When focus is on information already known or outside of what learner needs
Educational institutions like museums	Paulette, Paul, Liz, Tiki, Kristen, Warren, Ketan	Provide concrete, tactile experience Provide motivation Assist with historical information	When information is too vague or too detailed or irrelevant to need

(Continued)

TABLE 8.1 (Cont.)

Resource	*Used by*	*When useful*	*When not so useful*
Peers, friends, colleagues	Ana, Mary, Penny, Richard, Paulette, Kristen, Warren, Ketan	When have had similar experiences can guide, advise, support Access to unique knowledge and insight Cooperation	When the emphasis is on the barriers and difficulties rather than solutions and support
Clubs, associations	Mary, Richard, Michael, Paulette, Tina, Liz, Tiki, Warren	Social support and information to fulfil needs Sharing of problems to find solutions Sharing of resources; e.g. hardware too expensive for individuals to purchase, or prepared library of resources/ ideas	Need sufficient frequency of interaction to provide support
Personal resources	Richard, Michael, Liz, Tina, Paul, Tiki, Paulette, Kristen, Warren, Keith, Ketan, Hugh	Prior knowledge of how to seek information Ability to sort through information to find what is useful Persistence in trial and error Good social skills Range of personal skills, perhaps learned in previous employment	

Warren said he consumed 150–200 books a year, but he also used *"websites and Google when wanting quick updates. Then there are TV documentaries, museums, guide colleagues and, quite often, the guests themselves, as other sources of information"*. The case stories revealed that the usefulness of resources was often determined by the way our adult learners accessed them. In the next sections we explore these ways.

Many learning resources, including lectures, courses, exhibitions, books, television programs, information brochures, and so on, are essentially information givers. That information is available to be accessed by learners, but there is limited opportunity for the learner to offer a response in terms of how well it matches the information being sought. In schools and in other taught courses, for example, there is usually a set curriculum. Teachers have a responsibility to teach that curriculum regardless of its fit to learners' current learning needs. Although learning may occur, there is a definite sense of one-way communication. Further, it is very likely that a fixed curriculum or teacher-led course may not fulfil the learner's current needs. To increase the chance that it does, there needs to be two-way communication. The "teacher" has first to listen to the self-directed learner, and then decide what to do next to assist the process of learning and move the learner forward – a two-way communication. Resources are often more helpful if means are provided to enable interaction between the learner and the resources because this makes the communication two-way. If the resource is static, then very likely personal contact will be needed to help with that communication.

In the following sections, we discuss learning resources, particularly those listed in Table 8.1 in five clusters: media, including print, television, and Internet; people, including experts, friends, and peers; course-taking and teachers; visits to educational institutions; and personal resources.

Media Resources

Media are ubiquitous. Historically, the printed word has been the fundamental means of knowledge transfer and the basis upon which to build further knowledge. The advent of the Internet, the massive data storage capacity of the "cloud", and the mobile electronic devices used to access them, have combined to erode the predominance of the printed word as a source of information. The effects of the rapid expansion of electronic media are taken up in the next chapter. However, we recognise that our case story authors are adults and most of them grew up in a world where books were fundamental to learning and exploration. In fact, the number of books in a household has frequently been used as a component in measuring level of education in national studies of achievement, such as the National Assessment of Educational Progress (NAEP) in the United States (see, for example, Cowan et al., 2012) and international studies such as PISA (OECD, 2010).

Printed Resources

Many of our case authors began their search for information with books, indeed all of them used books to some extent. Richard relied on old books almost entirely for information about building his Baird televisor; Tina found a book about bees was a strong motivator during her participation in a citizen science project; Warren found books invaluable sources as he learned the background information necessary to become an effective guide. Of course, books were not always the best avenue for information. Ana found the doctor's book on embryology too technical and Michael found that his first books on making jewellery over-simplified the processes involved. Paulette's first incursion into books about geology was not encouraging due to unfamiliar terminology and use of jargon, but she found they became more useful as her knowledge progressed.

As well as books, print resources may take the form of newspapers, magazines, brochures, pamphlets, maps, graphs, diagrams, and other means of representation that aid understanding the messages to be conveyed. The form resources take depends on their purpose; for example, the intention may be to inform, such as a pamphlet on how to set up a worm farm or to give details about a particular medication; to encourage participation, such as travel brochures; or to entertain, like most magazines and fiction books. Some resources both inform and entertain, like the humorous book Tina found on bees. But all of these resources illustrate one-way communication. Sometimes this is enough; for example, Keith carries horticultural books in his car as an immediate reference about the different plants he comes across. Most encyclopaedias, dictionaries, and atlases fit into this category. Their usefulness is increased when the information they contain is well organised so it is easy to find. Dictionaries and encyclopedias are usually arranged alphabetically, and in other books generous tables of contents and indexes are essential aids to seeking specific information. "Way-finding" guides in the front of other reference volumes make their content more readily accessible to readers. Self-directed learners need to develop the skill of finding the information they want and how to make best use of it. Like most things, way-finding skills improve with familiarity and practice.

Broadcast Media

A common source of science information used by our authors was broadcast news and television documentaries. Ana followed some interesting debates in the press preceding a referendum to legalise abortion and found that embryological information was interspersed with a range of myths and misinformation. Ketan realised that not only did broadcast science fiction suggest the potential of science but many programs were also powerful conveyers of pseudoscience. What is the value of the science to be learned through media?

In 1922, John Reith, the BBC's General Manager, envisioned an independent British broadcaster able to educate, inform, and entertain (BBC, 2017). Dugan (2014) pointed out that "the century-old dictum on the need to educate, inform and entertain [actually] dates back to a statement made in 1922 by David Sarnoff, then head of the Radio Corporation of America, RCA" (p. 37). Dugan added that although "entertain" was the third word in this trilogy, the "practical experience of trying to reach an audience in an industrial, market-driven society soon made it clear that there was a need to entertain in order for the medium to achieve the other goals" (p. 37). The creators of mass media, of any kind, therefore find themselves forced to consider entertainment as a precursor to winning audience share, and concomitant risks in conveying science information vary from trivialisation and superficiality to misrepresentation.

Recently, in reporting on a discussion between media practitioners and science academics, Dugan (2014) noted how easily

> science can be misrepresented in popular media. Sometimes when there are disagreements between scientists, the controversies are glossed over to simplify the message. At other times the sparks of controversy are fanned into an inferno that generates more heat than light. But increasingly an understanding has been reached that science is not an activity that can be conducted in private solely for publication and discussion in academic

> journals. To obtain funding, appreciation and sympathy for science, scientists realise they need to engage the public through mass media.
>
> *(pp. 37–38)*

Dugan (2014) argued that scientists are willing to cooperate with media, and narrated ways he has cooperated with scientists to bring science stories to the screen. With a scientist "on board" the science has more chance of reliable interpretation.

Television, as media, competes both for funding and for audience, so presenting science as news or documentaries is often an uncomfortable compromise between what is "real science" and what is created to communicate it. For example, computer animation has developed to a stage where it is often impossible to tell whether what we are viewing is real or simulated. Paul's search for information about dinosaurs led him to the BBC's Walking With Dinosaurs, a high quality TV series first screened in 1999, in which prehistoric beasts were digitally created and portrayed in biological storylines. Rose (2001) claimed the episodes he saw "made me, and I suspect other biologists, squirm" (p. 115), because of "the inability of the programmes to distinguish between known fact from interpretation and sheer speculation" (p. 116). This was not a problem for Paul and Emma, who quickly found their DVD and book sources were out-of-date and therefore looked further afield for the information they sought. As Paul discovered, old TV documentaries become out-dated, but they live on packaged into digital resources. While the currency and relevance of their information may decrease as science moves on, their entertainment value may remain.

The conflation of education and entertainment in providing science education via mass media remains a problem. Although Rose (2001) made the following suggestions two decades ago (to a BBC Governors' meeting in 1999), his ideas to improve science broadcasting in the 21st century remain relevant. Rose suggested the following dot points:

> - recognise that there are many sciences;
> - show the links between science and technology;
> - mitigate the gender gap;
> - recognise that scientists are not disinterested;
> - go beyond press releases;
> - experiment with turning listeners and viewers into scientists;
> - expose scientific claims to critical analysis; and
> - open up the debates and uncertainties within the sciences themselves.
>
> *(Rose, 2001, p. 119)*

Unfortunately, broadcasters do not routinely follow Rose's suggestions, thus denying learners access to reliable science information, either by simplification, misinterpretation, or obfuscation. Learners are left to judge the credibility of their offerings, a point we return to in the next chapter.

The Internet and Social Media

We devote the next chapter to new media; here we simply note that our authors searched online for useful websites (online courses are mentioned later in this chapter). Ketan, in his pursuit of pseudoscience and trying to understand *"why people believe weird things"*, was a prolific user of media, especially the Internet. Here, too, the credibility of information can be an issue and our authors were all were aware of the need for selective use; Tiki specifically warned of the necessity to cross-reference the information found. With smart phones, learners have a mobile source of information in their hands, so the use of online resources is becoming more prevalent (see next chapter). Further, and unlike the one-way communication of the Internet, interactive media offer avenues for two-way communication. Some of our learners referred to their use of social media, a new but pervasive information source for some people. But as our authors pointed out, careful vetting is required as social media is a persistent purveyor of opinion rather than validated information, a caution that must be extended to the Internet (Brossard & Scheufele, 2013).

Quality of Information Portrayed by Mass Media

Determining the quality of data available from mass media can be troublesome for self-directed learners searching for information pertinent to their need. While information available online and in printed brochures from health, environmental, or government resources should be up-to-date and accurate in the sense of being based on the best available evidence, information in popular magazines and commercial or social websites is often less reliable. Wilson, Bonevski, Jones, and Henry (2009) evaluated 1,230 health news stories over 4-year period from commercial and publically-funded newspapers, online news, and transcripts from broadcasts on radio and television. Each story was reviewed against ten criteria and the percentage of "satisfactory" ratings calculated. Broadsheet newspapers were found to be the best performers, but their average score was only 52%. Current affairs TV programs scored lowest at 33%. More recently, Wilson (2016) used the same set of criteria to evaluate ten popular (by circulation rates) Australian-based magazines over a six-month period. Those with "health" in their title were the worst performers, the others performed poorly, but one magazine, *Dolly*, had a perfect score on all of its stories, attributable to their being written by general practitioner who is also a medical researcher. The message to be learned is two-fold. First, many of the people responsible for purveying information need to learn how to do it clearly and accurately, and second, readers must learn to pay careful attention to the source of information, and be able to judge whether or not the information is evidence-based.

People as Resources – Experts, Friends, Peers, and Colleagues

Interacting with other people is the archetypical means of two-way communication. We live a social existence and we learn by interacting with others, so it is not surprising that our authors frequently referred to their experiences of learning from and with other people. Sometimes these people were experts in their field, sometimes they were personal friends or family members, and other times they were members of a club or association devoted to the field of interest to our authors. Some of our authors, like Hugh and Keith, benefited greatly from being mentored by more senior or experienced people in their work place. Indeed, successful "learning on the job" invariably involves mentoring by others. In this section, we first look at experts in the field, beginning with medical experts.

Experts in the Field

In Chapter 3, Ana, Penny, and Mary, three of our case story authors, related their experiences about a particular health issue, and although all of the medical experts with whom they interacted were willing to assist, invariably there were gaps in communication that forced them to search elsewhere for the information they needed to deal with their issue. Ana found that being alert to learning opportunities, such as reading the details on an amniocentesis permission form and talking to a nurse after a antenatal course, enabled her to make progress in understanding her pregnancy. In Penny's case, her symptoms were varied and complex and building up an information base to reach a diagnosis and therefore consider treatment has so far been unsatisfactory. Having a Down Syndrome daughter resulted in Mary's need for continuous self-directed learning, and action on Mary's behalf for the benefit of her daughter.

During our lifetimes we will all need to seek advice from health professionals and we know it can be tricky. Ha and Longnecker (2010) reviewed research relating to doctor–patient communication and identified some of the barriers to effective communication. These included "patients' anxiety and fear, doctors' burden of work, fear of litigation, fear of physical or verbal abuse, and unrealistic patient expectations" (p. 39). They recommended communication training for doctors, and emphasised the importance of collaborative, two-way communication between doctors and patients.

Advice about how doctors can best communicate with their patients is a common theme of literature in the health professions, but while advice for physicians is easy to find, advice for the patients is not. As Ana, Penny, and Mary quickly discovered, learners need to be persistent and know how to ask the "right" questions. The US-based National Patient Safety Foundation (NPSF) requested health literacy experts to develop an educational program that would improve communications between health care professionals and patients and their families. The result was the "Ask Me 3®" educational resources (NPSF, n.d.).

These are available for the use of the professionals to assist them to encourage their patients to ask three specific questions: What is my main problem? What do I need to do? and Why is it important for me to do this? The intention is to help patients become more actively involved in their health care.

The University of California Medical Center (UCSF, 2002–2017) also offers a list of advice to patients to help build the doctor–patient partnership. The list is very comprehensive and requires considerable activity by the patient, including thorough preparation before the visit, planning questions, sharing needs and concerns, and taking notes during the visit. Although this list is aimed at cancer patients who likely have an ongoing relationship with their health care professionals, the general advice of "balance assertiveness with friendliness and understanding" is probably something that works both ways.

The thrust of this discussion resides in two-way communication. Patients need to tell their medical expert about their problem, the expert has to listen and decide on advice. But to do this well, both people need the basic skills to enable them to focus the discussion on the points that are salient in the context of the situation; the patient's general health and lifestyle, for example, and the treatments that are available. This same general principle applies in communication between any kind of expert and learner. Sometimes the context is complex; for example, Paulette's attempts to seek information about the local water supply from employees of the water company were compromised by the employees' need to protect their employer's interests. In contrast, Richard's attempts to find out the working of old technological components from experts, the technicians who had actually used the components, were successful because he knew what he needed to know and the technicians enjoyed sharing their experience and the feeling that their advice was valued.

Learning from experts, in any field, depends not only on their willingness to assist, but their ability to assist. The learner needs clear and accessible information, so the expert needs to understand where the learner is "at" and what kind of knowledge will best match the learner's level of understanding. Of course this requires clear exposition from the learners, who must be able to articulate their requirements, and to do this they may need help from the expert, in the way of thoughtful questions to help focus their need. However, many experts do not learn the skills required for effective communication as part of their job training, so there is often a hiatus between them and those who seek to learn from them.

Friends, Peers, and Colleagues

Friends, peers, and colleagues are not necessarily different people. Many of our authors exchanged ideas and experiences with individual acquaintances dealing with similar issues, such as Paulette's friends who helped to monitor the health of local waterways, and Kristen's science centre colleagues who were a significant source of information as she learned to become an explainer. Like Kristen, several of our authors were mentored "on the job" by more experienced colleagues. Hugh and Keith were mentored by others in leadership positions; Tiki and Warren learned not only how to become guides but also more about the context of plants or local history, as they were mentored by other, more experienced guides. Sometimes friends can be helpful by simply serving as a sympathetic "sounding board" as learners talk about their attempts to solve problems or pursue the information they are seeking. This helpful role assists learners to rehearse what they know and identify what additional knowledge they need.

Other friends and colleagues may be members of a club or association formed to further particular interests of a group of people. Mary, for example, received considerable benefit by interacting with other people who had Down Syndrome children and she later became a leader in her local association. Richard and Michael were active participants in clubs that facilitated social interaction relating to old sound technology and silver-smithing, respectively. It is common for hobbyists to belong to associations that enable social and educational interaction relating to their hobby. Corin, Jones, Andre, Childers, and Stevens (2017) describe how science hobbyists (in this case birders and amateur astronomers) "reported that their relationships with [their hobby] organisations, and the people they met through the organisations, were invaluable in getting them more involved in their hobby" (p. 175). Richard and Michael were both founders of their particular clubs, but through the clubs, they met a range of other people from whom they could learn and also, less experienced members whom they could mentor. Warren also formed a group with other interpreters and they consistently shared their knowledge. Mary's participation in her local Down Syndrome organisation enabled her to assist others, not only by being secretary, but also pushing for the acceptance of Down Syndrome children in mainstream schooling.

Tina's bee walks were part of her participation in a citizen science project to monitor Britain's bee population. Citizen science, sometimes known as public participation in scientific research (Bonney et al., 2009), refers to public science-related programs that involve the voluntary contribution of data by ordinary citizens. Most citizen science projects are contributory, but others are collaborative, where the public helps with design, data analysis, or disseminating findings. A few are co-created, where at least some members of the public are involved in all steps of the research process. The citizen science movement is amazingly widespread, with projects in almost every field of science. Liz was involved in an international online citizen science project called Galaxy Explorer, in which people are recruited to help classify pictures of galaxies. Essentially, such programs work to enable to scientists to collect and screen data on a much greater scale that could be done by themselves. While both Tina and Liz, and other learners, enjoyed their participation in such projects, there is increasing evidence that participants gain knowledge about science and its processes and such projects also increase public awareness of science and the diversity of research (Bonney, Phillips, Ballard, & Enck, 2016). For individual learners looking to further their knowledge about some aspect of science, participation in a citizen science project is a rewarding way to learn and experience a sense of contribution.

Course-Taking and Teachers

Taking a course is a structured means of learning for one's self and there is an almost limitless array of courses available for the casual learner or those who wish to further their professional learning. The Open University in the United Kingdom has been accepting students since 1971 and is now one of the world's largest universities. The Open University has always been groundbreaking in the use of technology to deliver its courses via distance education, making them accessible to any person in the world. There are numerous vocational and training colleges offering post-school opportunities for learning, and more recently, a range of learning opportunities have become available via the Internet. However, many formal courses, because they have a definite length, will usually contain more material on a given topic than the self-regulated learner will want or need. It may therefore be the case that the self-regulated learner, having satisfied their initial needs in other ways, will not be attracted to take such a formal course.

Four of our case study authors chose to take a course that they thought would provide the kinds of knowledge and skills they required to continue their learning journey. These courses were very different and served different purposes for our authors. Hugh was the most prolific course taker as he worked his way to higher levels in his employment. His technical courses were structured to cover the specific content and skills prerequisite for continuing employment or promotion within the aviation industry. For Hugh, the motivation to take these courses was simply a way to maintain the challenge in his job, but when he found that he needed more knowledge about the insurance aspects of his work, he took a different approach: *"Guidance and basic training was given by my employers, however I took it upon myself to read up on Insurance Industry rules, regulations and protocols from various publications and training documents handed down to me from others."*

At the appropriate time during her pregnancy, Ana attended an antenatal course. This was a formal course structured around the stages of pregnancy and designed to prepare mothers for the process of birth and preliminary post-natal care for their newborn child. Perhaps because Ana had been proactive in seeking such information herself, she found the course too basic for her needs. What she did find very useful were the post-session discussions she was able to have with the obstetrics nurse who ran the course, so Ana attended each week for the opportunity to have these discussions as her pregnancy progressed and new questions became relevant to her. Somewhat like Ana, the benefit of Michael's technical course was dependent upon a supportive instructor. Michael decided to take a course because his efforts at jewellery making were not succeeding.

> *I knew I needed help. I found that the local technical college was running evening classes in silversmithing. I went along to the open evening to see if it was what I wanted. I thought it was so I registered and for the next few years became a student.*

Michael's found his course-taking very useful because instead of listening to formal instruction, students had the flexibility to pursue their own projects assisted by the tutor who was a well-trained silversmith. Repeating the course in subsequent years did not entail repetition of content, rather it provided opportunity to attempt more difficult projects with an expert guide and access to specialised equipment that Michael could not afford. When the course closed, Michael and some of his fellow students formed a club, purchased the advanced equipment they needed to continue their jewellery making, and supported each other in their projects.

By chance, Liz came across a "science for mothers" course at a time when she was wondering how she could assist her children with their science homework but feeling she lacked the necessary skills. Although it was a structured course, the female tutors were very flexible and Liz found it to be a safe environment in which to learn. The course gave her confidence to seek further science learning opportunities, and she scoured the Internet for consolidating activities. She completed a course on Kitchen Chemistry and eventually branched out into some citizen science projects on wider topics.

Other types of courses typically accessed by self-directed learners are the brief training courses offered for volunteer activities. Tina's interest in bees led her to search for information about bee identification. She found *"a day's training each year proved to be far too little but it clarified the main aspects that identification involved"*, and then pursued other sources of information to fill out this general framework. Tina wanted to ensure her bee identifications were accurate because she was concerned she might inadvertently provide inaccurate information to the bee data base to which her "bee-walks" were contributing. To some extent, Keith's needs were similar. He often needed a plant identified or some information to rescue a plant that was not thriving. Both Tina and Keith needed information quickly so that they could take action to correct a misidentification or tend to an ailing plant and move on. Keith has considered further study: *"I am looking at further study, though, because there are gaps. I want to know more – I want to know why things happen."* However, it is unlikely a generally structured course would suit his need for on-the-spot information. Tina's bee-learning experiences have informed her self-knowledge about the way she prefers to learn, and how she will address her new interest in plant identification.

> *I believe I am more able to tackle this new challenge more effectively. For example, I now know that going on a generalised course, in this case on flower identification, does not suit me. It would provide too much knowledge, much of which is not pertinent to*

> *what I want to know, and would be overwhelming. Consequently, I use a "need-to-know" strategy in which I only focus on flowers that bees visit and ignore other flowers.*

Tiki, Kristen, and Warren, as interpreters at a botanic garden, a science centre, and heritage sites, respectively, also had sequences of training courses to provide the science background needed for their interpretative work. Apart from the personal study skills necessary to take in large amounts of information, they also learned in a very social way from the scientists and the other interpreters or guides around them, so their learning was not limited to content, it also involved the social skills of guiding others. Tiki emphasised the importance of learning in context, and also the importance of being able to listen and learn from others.

> *There was a fair emphasis on interesting facts, because those would be of regular use to us, but every fact had to be set in its context at least for our understanding so that we could use it correctly. I still have the training materials from that course, because they make an excellent refresher whenever I can't remember how this or that bit goes. The course itself was steady in its progress, presented by a mix of expert plant people, expert communicators and experienced guides. Through the sessions we had a lot of access to good people, and the same attentive listening skills I'd developed for radio came into play.*

The experiences of our case study authors illuminate the usefulness of course-taking for the self-directed learner, but usefulness depends on the match between what the participant needs to know and what the course is structured to provide. Narrowly focused courses such as those taken by Hugh are effective for those wishing to advance their work-place qualifications, but they need to be completed. Warren, while still at school, completed night classes to obtain a ham-radio operator's licence. Self-directed learners have the motivation and self-regulating skills that enable completion, even if they find some of the course not relevant at the time or even tedious. Other courses less formally structured can give learners the opportunities to interact with like-minded others. For example, both Tina and Michael were able to meet people with the same interest and discuss their progress with them. In Ana's antenatal course and Michael's jewellery-making course, a major benefit was the support of the tutor who was willing to be sufficiently flexible to answer questions and provide guidance to learners who know what they want to know.

Internet e-Learning Platforms

The internet environment enables a range of e-learning platforms, as Liz found when she discovered a course on Kitchen Chemistry in her search for more science. Online courses abound; they range from short (one minute or even less!) tailored lectures/demonstrations on very specific topics to lengthy series of sessions. Indeed, it is difficult to find an everyday topic that does not have at least one YouTube video to explain or demonstrate it!

What were previously "distance" or "correspondence" courses, where students do not attend classes, are increasingly being offered online. An innovation in 2008 was MOOCs – massive open online courses – that are offered via the web. Usually free and open access, MOOCs can have unlimited participants from any geographic location. The courses are usually taken informally, but some are structured and certificated. There are thousands of MOOCs; one list of MOOCs is available from https://www.mooc-list.com/. On the day of writing this sentence, 15 different science courses were scheduled to begin. They were in English, Spanish or Chinese, and from a range of providers, mostly universities, but opportunistic commercial organisations are also offering MOOCs. Mostly MOOCs are one-way communications of varying quality and interest, ranging from a set of lecture notes to complex, multimedia platforms with which learners can engage interactively. Increasingly, some MOOCs aimed at certification have teaching assistants and/or user forums, like chat rooms, to support interaction among participants in the course, the course writers, and the tutors.

MOOCs offer a flexible learning environment in which participants can choose when and how they engage, and the level of effort expended. They are ideal for self-directed learners to engage in topics that complement their professional employment in both a formal and informal way, or to simply access information that is of interest to their hobby or their health. Completing a course independently requires a considerable degree of self-regulation, and this is more likely to be found in participants who see instrumental value in the course, either for their workplace or furthering their interest, rather than participants who are focused on obtaining a certificate for course completion. For example, Hood, Littlejohn, and Milligan (2015) and Littlejohn, Hood, Milligan, and Mustain (2016) examined self-regulated learning in an online course, "Introduction to Data Science", offered by the University of Washington. At the time of their research the MOOC had attracted 50,000 participants from 197 countries. Using a combination of survey and follow-up interviews with data professionals who completed the course, Littlejohn et al. (2016) reported "Distinctions were detected between formal, certification-oriented participation where learning was primarily centred on the MOOC versus non-formal, professional-development oriented participation where learning was centred on the utilisation of knowledge in real-world contexts" (p. 47). Further, Littlejohn et al. found that for highly self-regulated participants, "Their focus on expertise development for their professional roles rather than formal assessment translated into higher overall satisfaction" (p. 46). Hood et al. (2015) remarked on higher levels of self-efficacy among the data professionals and those seeking a higher education qualification compared with others taking this MOOC. These authors concluded that this finding "highlights the need for greater attention to be paid to how MOOCs can better support learners from all backgrounds" (p. 90).

In sum, then, the successful use of MOOCs as a learning resource is dependent on the motivation and self-regulating skills of the participant. Given that most MOOCs offer one-way communication and have no access to a tutor, the learner must pick and choose to find what is useful among the offerings. Like the other courses accessed by our case study authors, MOOCs are likely to be successful when they meet the needs of the participant, and the learners have the internet skills to access and pursue the course. We can turn around the old adage "horses for courses", where one chooses the horse to match the condition of the race course. For our self-directed learners, it is a matter of "courses for horses", where they can choose the course that matches their needs.

Self-Directed Learning at Education Institutions

Libraries, museums, science centres, botanic gardens, and other educational similar institutions are characterised as free-choice learning environments; visitors can choose what they want to do and where they wish to go. Banz (2008) summarised some of the literature on self-directed learning and its implications for museums, pointing out the significance of the free-choice environment. Because visitors can choose how to interpret the available information, learning was very personal for each learner. This is reflective of Falk and Dierking's (2000) Contextual Model of Learning that attributes the learning experience to three contexts relating to the personal qualities of visitors, the physical environment, and the social context, which includes who visitors are with and with whom they interact.

Of course, adult visitors are not necessarily visiting for their own learning. Liz loved to take her children to the local science centre and her case story suggests that she enjoyed the opportunities for learning herself. Paul was familiar with the local natural history museum so he took his daughter for the purpose of learning about dinosaurs for her school project. As is often the case, Emma was distracted by other exhibits but they eventually reached the dinosaurs and achieved their goal. Paul remarked *"I wonder, if it had not been for this museum visit, I might have felt intimidated by the vastness of the subject matter and would have relied solely on the Internet for the quickest of kills"*. It seems that the familiar "concreteness" of the exhibits focused their search for project information and acted rather like a sorting mechanism for the vast subject matter on dinosaurs, much of it online and less tangible.

Sorting through vast subject matter is a task for those who work as interpreters or guides, like Tiki, Kristen, and Warren. Tiki was a volunteer interpreter at a botanic garden, learning about plants in a way that Orr (2006) has described as "serious leisure". Orr noted that volunteering serves to provide "people with the opportunity to learn and participate in the social world associated with their chosen field of interest" (p. 210). Volunteering as an interpreter requires training in the field, something that Tiki enjoyed because she was building on her early deep interest in plants, and she put considerable effort into getting the botany right because she was helping others.

Kristen's career move to an explainer role at a science centre was motivated initially by her desire to interact with people, rather than a love of science. She had a steep learning curve because of her limited background in science but, she said, *"from the start I had confidence that I could do this. I can research, I can study, I can relate to people"*. For Kristen, the science centre was a resource for learning as well as her employment. In contrast, Warren's decision to become an interpreter was a natural consequence of his interest in conservation and where he lived.

> *I took the concept of really knowing about the place where I lived to heart and started to learn more and more about Dunedin, the Otago region, and southern New Zealand. I wanted to be very knowledgeable about my part of the world.*

Warren was a frequent visitor to the local library and museum, learning about local Maori and European history that seems related more to heritage than science, but natural history – flora and fauna – is very much part of heritage and science.

Paulette, herself a heritage consultant, had no hesitation in looking to the local museum (which was also the town library) for assistance in developing her understanding of the local geology and its effect on Mere's water supply. With the help of the curator, she sought historical data about water use to establish baseline data about water levels, and she was able to use the museum as a collection point for her survey of residents' water stories. Paulette's strong research skills were a great help in enabling her to gather the data, organise the information, and build the three-dimensional picture she needed to understand the fluctuating water levels in the local chalk streams. Her knowledge grew to the extent that she now leads U3A groups in geologic tours along the Mere Fault.

Banz (2008) noted that self-directed learners formed a large part of museum audiences and would "continue to benefit from the rich educational opportunities of the museum as long as their self direction is supported" (p. 52). Our case story authors were self-directed learners who chose their visit destinations freely, but they were not "window shoppers", a term sometimes used for browsers in museums (who also form a large part of their audience), they were there for a purpose, such as Paul's visit to study the dinosaurs. Whether or not their purpose was achieved was determined by their familiarity with the institution and its staff, and what they had to offer. Guides, like Tiki, Kristen, and Warren, not only need to learn and understand their subject(s), they need to know how to interpret and communicate the information to others.

In this chapter we have seen the benefits of two-way communication, an interchange between a learning resource and the self-directed learner seeking help. Often the resources available in educational institutions, like museums, are exhibits or other forms of information that can only communicate one-way, and while that can be a perfectly satisfactory source of information, often two-way communication is more useful because interaction with the learner can focus the search and reduce the time sorting through irrelevant information or pursuing "dead-ends". There is much museum and other guides can do to provide this support, and this will be taken up in Chapter 10.

Personal Resources

The last row of Table 8.1 refers to personal resources, personal attributes that our case study authors drew upon to help them on their learning journey. Naturally the attributes that were most important varied according to the nature of the task. For those seeking medical information, finding the right questions to ask of the right resources and filtering through often quite complex information were essential to progress. For Richard, it was his knowledge of electricity as a physics teacher and his ability to experiment and, through trial and error, find "work-arounds" when he could not obtain original parts for his Baird televisor. Kristen's wish to communicate with other people and her social skills were a fundamental personal resource as she began to build her science background "on the job" of being an explainer at a science centre.

There were, however, several tasks our authors had in common as they progressed in their journey. Each of our authors had to find the appropriate resources – one or more of the many kinds of media; experts, other people, friends and peers; courses or teachers; other educational institutions, that could provide the information they needed. The next task was to access those resources, coping with the variation in cognitive level, jargon, or expected level of skills, to extract the information. Then, to make best use of that information, judgements had to be made about its credibility and then ways to organise it so it was meaningful. Making a map of Mere's water sources was a turning point in Paulette's understanding of the Mere Fault, for example. And through all of these tasks, our case story authors had to determine whether or not they were actually making progress towards their goal, because if they were not, then a new direction, another resource, was required. This self-monitoring can be a learning step in its own right. By reflecting on their own learning, our authors increased their understanding of how they learned, something demonstrated clearly by the Moon Diary assignment. Understanding how they learned streamlined subsequent learning steps because learners' experience showed what kinds of resources were most likely to be helpful to them. Tina's reflection on her learning shows this admirably:

- *There is an initial ad hoc approach to learning because the learner doesn't know enough about the subject to learn in a logical fashion and does not know where to get help. The learner could give up at this stage.*
- *Total self-study needs a lot of self-discipline as other life demands can easily take over time. However, volunteering in a group provides structure, some feedback, and motivation to continue.*

- *It is important to take full opportunity of increasing on-line support and chance to meet other interested learners.*
- *Expect to take a step-by-step learning process: I found a "need-to-know" strategy suits me.*

Tina's first dot point is a very significant one. It is very easy for a learner to give up early in their journey when they lose direction and do not know where to turn for the resources they need. Our self-directed learners were persistent, they did not give up. Even when Penny was unable to get a firm diagnosis for her medical problems, she came to terms with the situation, but was ready to pursue further information if her condition changed. The underpinning reasons for their persistence were three-fold: our learners were motivated to achieve their task, they knew how to actively engage with resources they used, and they had positive self-efficacy, in that they had confidence that they would succeed. These three important personal characteristics of self-directed learners will be explored more fully in Chapter 10, after we deal with the last cluster of resources, the Internet and new media, in the next chapter.

References

Banz, R. (2008). Self-directed learning: Its implications for museums. *Journal of Museum Education*, 33(1), 43–54.

BBC. (2017). The BBC Story 1920s [Fact sheet]. Retrieved from http://www.bbc.co.uk/historyofthebbc/resources/fact-sheets

Bonney, R., Ballard, H., Jordan, R., McCallie, E., Phillips, T., Shirk, J., & Wilderman, D. (2009). *Public participation in scientific research: Defining the field and assessing its potential for informal science education (A CAISE Inquiry Group Report)*. Washington, DC: Center for Advancement of Informal Science Education (CAISE).

Bonney, R., Phillips, T. B., Ballard, H. L., & Enck, J. W. (2016). Can citizen science enhance public understanding of science? *Public Understanding of Science*, 25(1), 2–16.

Brossard, D., & Scheufele, D. A. (2013). Science, new media, and the public. *Science*, 339, 40–41.

Corin, E. N., Jones, M. G., Andre, T., Childers, G. M., & Stevens, V. (2017). Science hobbyists: Active users of the science-learning ecosystem. *International Journal of Science Education, Part B*, 7(2), 161–180.

Cowan, C. D., Hauser, R. M., Kominski, R. A., Levin, H. M., Lucas, S. R., Morgan, S. L., … Chapman, C. (Expert Panel). (2012, November). *Improving the measurement of socio-economic status for the National Assessment of Educational Progress: A theoretical foundation (Recommendations to the National Center for Education Statistics)*. Retrieved from https://nces.ed.gov/nationsreportcard/pdf/researchcenter/Socioeconomic_Factors.pdf

Dugan, D. (2014). Science story telling in TV documentaries. *Actes d'Història de la Ciència i de la Tècnica, Nova Època* (in English), 7, 35–49. ISSN: 2013–9640 / DOI: doi:10.2436/20.2006.01.183. Retrieved from http://revistes.iec.cat/index.php/AHCT

Falk, J. H., & Dierking, L. D. (2000). *Learning from museums: Visitor experiences and the making of meaning*. Walnut Creek, CA: Altamira Press.

Gardner, H. (1993). *Multiple intelligences: The theory in practice*. New York: Basic Books.

Ha, J. F., & Longnecker, N. (2010). Doctor–patient communication: A review. *The Ochsner Journal*, 10, 38–43.

Hood, N., Littlejohn, A., Milligan, C. (2015). Context counts: How learners' contexts influence learning in a MOOC. *Computers & Education*, 91, 83–91.

Littlejohn, A., Hood, N., Milligan, C., & Mustain, P. (2016). Learning in MOOCs: Motivations and self-regulated learning. *The Internet and Higher Education*, 29, 40–48.

National Patient Safety Foundation (NPSF). (n.d.). *Ask Me 3: Good questions for your good health*. Retrieved from http://www.npsf.org/?page=askme3

OECD. (2010). *PISA 2009 Results: Overcoming social background – Equity in learning opportunities and outcomes* (Volume II). Retrieved from http://dx.doi.org/10.1787/9789264091504-en

Orr, N. (2006). Museum volunteering: Heritage as "serious leisure". *International Journal of Heritage Studies*, 12(2), 194–210. http://dx.doi.org/10.1080/13527250500496169

Rose, S. (2001). What sort of science broadcasting do we want for the 21st century? *Science and Culture*, 10(1), 113–119.

UCSF Medical Center (2002–2017). Communicating with your doctor. Retrieved 2017, from https://www.ucsfhealth.org/education/communicating_with_your_doctor/

Wilson, A. (2016). Looking for health advice? Don't consult health magazines, try Dolly. *The Conversation*, November 21. Retrieved from http://theconversation.com/looking-for-health-advice-don't-consult-health-magazines-try-dolly-63823

Wilson, A., Bonevski, B., Jones, A., & Henry, D. (2009). Media reporting of health interventions: Signs of improvement, but major problems persist. *PLoS ONE*, 4(3), e4831. DOI: doi:10.1371/journal.pone.0004831. Retrieved from http://journals.plos.org/plosone/article?id=10.1371/journal.pone.0004831

9

LEARNING FROM NEW MEDIA

New media – media platforms, technologies, and systems that have emerged in the Internet era – have profoundly altered the ways that adults learn and explore the world of science and technology. As many have observed, the ways that adults learned a few decades ago are radically different from the ways they now seek, find, and explore new topics. In this chapter, we will discuss how people learn about science and technology using these new information channels and how the people in our case stories approached this challenge.

While new media have changed the ways people can learn about science, this chapter also explores the continuities – the ways in which Internet-enabled learning rests on and connects with much older learning methods. In this light, the Internet is just one (admittedly powerful) tool among many. It is clearly impossible to write an up-to-date assessment of learning through the Internet, since this is changing even as we write. Rather, we seek to give an overview of the characteristics of such learning that are unlikely to change, and to tie these in with the stories in this book. Although the people in our case stories turned to the Internet for specific information, we have included a short section on social networks to illustrate some of the ways in which these new media are being used to communicate science.

This chapter also addresses the ways that learning on the Internet can be flawed, perhaps even dangerously flawed. We examine how poor search and critical thinking skills, cognitive biases, and the very nature of the social and algorithmic media that make up so much of our news and learning diet can be biased towards problematic learning. Finally, we suggest ways in which people can be encouraged to access the Internet with confidence to find out what they need to know.

DOI: 10.4324/9780429434303-9

Characteristics of New Media

The first observation that can be noted about new media is the historically rapid increase in users and connections. Some 88% of Americans used the Internet in 2016 (Smith, 2017). This number is echoed, for example, by 86% of adults in the United Kingdom (Office of National Statistics UK, 2015) and 86% of Australian households in 2014–2015 (Australian Bureau of Statistics, 2016), but it has been the speed of uptake that stands out. Just seven years after its launch in 1991, the world-wide-web was used by a quarter of Americans – it took 46 years for electricity to reach this figure, and 26 years for television (see Figure 9.1).

More recent Internet-focused technologies, such as broadband connections to the Internet, social media, smart phones, and tablets, have seen even faster growth. Tablets, for example, were used by one quarter of the population of the United States less than 3 years after their launch in 2010 (Smith, 2017). It is likely that such trends will continue in the future.

Second, we can also observe a rapid historical increase in raw data bandwidth and associated content richness. In straight technical terms, while the number of global Internet users has been growing at just under 3% per annum (International Telecommunications Union, 2015), global bandwidth has grown at a far faster rate – from 30 Tbit/s to 185 Tbit/s between 2008 and 2016 – an average of 19% growth per annum (International Telecommunications Union, 2016). This rapid increase in speed of data transfer has enabled ever more data-dense forms of communication. While the earliest forms of Internet-enabled communication were exclusively text based – email and bulletin boards – later years have seen the gradual emergence of richer forms of content. Images, audio, and video are now ubiquitous. Other new forms (such as virtual and augmented reality), while still in their infancy, beckon on the horizon. Stated simply, more and more of us are sharing more and more information, more and more content – and more and more of our lives – online.

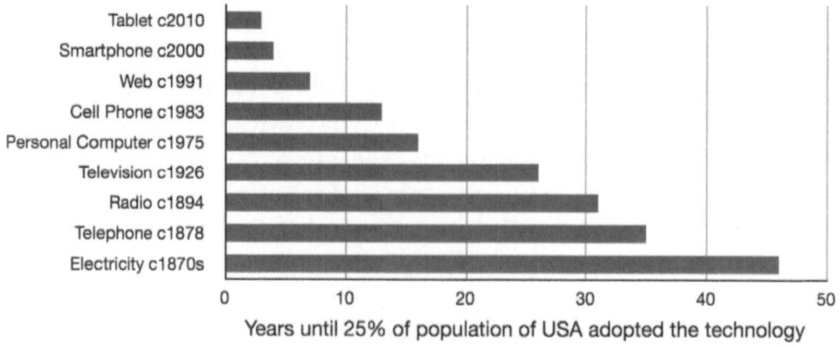

FIGURE 9.1 Rate of adoption of communication technologies in the United States

In the development of greater bandwidth and richer content it is important to note that more complex forms of communication have not simply replaced – and are not likely to replace – less complex forms. Here we can follow Goody's argument that "the written word does not replace speech, any more than speech replaces gesture" (Goody, 1977, p. 15). Email, despite the consternation with which it is viewed by modern office workers, is unlikely to disappear in the short or medium time frames, and is very unlikely to be replaced by some sort of video conferencing system. Video shared on YouTube has not made text-based communication redundant.

Further, the gradual development of richer forms of content does not preclude the subsequent emergence and success of what might be perceived as simpler communication forms, for example the development of text messaging/short messaging services (SMS) on mobile phones long after the development of more information-dense voice-to-voice telephony. Indeed, one can recall the dismally inaccurate predictions of failure that greeted the arrival of text messages ("why would I want to write something when I can use my voice?").

A corollary to this point is that while video has not replaced text communication, the arrival of the Internet has ushered in radical changes in written communications through the introduction of social media and text messaging. Physically printed media have seen significant upheavals in recent years, resulting in changes in media economics, behaviour, ethics, and governance.

The third observation concerns the massive growth in both numbers of users and bandwidth. We have seen a shift in the production of media from exclusively professional organisations – professional publishing houses, print media corporations, television stations – to a world that draws from and combines amateur, professional, and semi-professional content. Here hobbyists, interested amateurs, and those hoping to turn their hobbies into professions can upload a range of different media (text, video, audio, and imagery) to a range of different platforms and sit them right alongside content produced by these professional organisations. For its more enthusiastic advocates this shift towards more amateur – and hence potentially more open and democratic – media has offered revolutionary potential (see, for example, Shirky, 2008). Yet the revolutions brought about by this shift have not all been as democratic as many commentators might have hoped. The emergence of "fake news" is a clear example. Regardless, the development of amateur media via the Internet has offered a fascinating space for both discussion of a great variety of topics of interest, and the opportunity to "play" with them – a form, one can suggest, of active learning.

Learning Via the Internet: The Digital Divide

How have new media affected how adults have learned about science and technology? Despite the increase in Internet users worldwide, there remains a "digital divide", but the nature of this divide differs across studies. It may be a division of age, of socio-economic factors, of gender, of culture, or of literacy and skills – all these affect motivation and ability to use new media. In the United Kingdom, a major government initiative led to the establishment of 7,000 centres in places like libraries and community centres that afforded Internet access in order to "provide technology-mediated opportunities for learning .... These initiatives aim to widen participation in those groups traditionally underrepresented in adult learning" (Cook & Light, 2006, p. 51). It should be noted that facilitating adult learning, in this context, implies enabling enrolment in online courses. Gorard and Selwyn (2005) explained that the underrepresented groups "include women, the elderly, some ethnic minorities, those on low incomes, ex-offenders and people with learning difficulties" (p. 72). These authors quoted the official rationale for setting up these centres as having "the potential to radically improve participation and achievement rates in education". Benefits were stated to include individual, customised learning and flexibility, both of time and place. Cautioning against over-optimism, however, Cook and Light (2006) stated that "at a minimum, a citizen needs to be able to, and feel confident in their own ability to, use a search engine and to send an email before they can make use of e-government services" (p. 52).

In a key paper seeking to assess the impact of these initiatives, Eynon and Helsper (2011) described a survey of 2,350 people in Britain to ascertain the level and nature of Internet use. They found that most people actually accessed it from home, not from external centres. They stated that "while 39 per cent of adults have undertaken some kind of learning activity in the past three years, only 18 per cent are currently learning" (Tuckett & Aldridge, 2009, as cited by Eynon & Helsper, 2011, p. 535). In this context, once again, "learning" applies to formal learning activities. In their survey however, Eylon and Helsper distinguished between online formal learning, and informal learning, which included "finding information about health and medical care" (p. 539).

In an earlier report, Selwyn and Gorard (2004) had also questioned then current wisdom about learning from the Internet. Much like the case stories we have discussed, adults in their study accessed the Internet for many reasons other than undertaking formal education in this informal environment. Of 1,101 respondents in the UK, most were not engaged in learning in a formal sense. Of those who were learning, most were using the Internet to further work-related knowledge and skills (p. 304). There were also activities related to leisure and general interest. "However, it should be noted that these ICT-based episodes of intentional informal learning were usually part of an ongoing sustained project that also involved learning via books, courses and social contacts" (p. 304). This mirrors the narratives in our case stories.

In terms of helping people, Eynon and Helsper (2011, p. 548) concluded that, instead of focusing on formal online learning,

> supporting informal learning may well be the most effective course of action. Informal learning is an extremely important part of individuals' learning experience and can have all kinds of social and, in some cases, economic

> benefits. Furthermore, increasing engagement in informal learning online could potentially awaken an interest in more formal, certified types of learning among those with educational disadvantages.

Gorard and Selwyn (2005), expanding on their earlier (2004) report, also found that most participants used the computer for learning, both formally and informally, at home rather than in an outside environment. Of course the introduction of new media such as smartphones means that people can access the Internet almost anywhere, and some science learning will surely be done through this avenue. A key finding of this study, however, was that the Internet itself did not appear to be "a key participant in adult participation in formal learning". Rather, "those reporting a sustained period of self-directed study or leisure learning were nearly twice as likely to be lifelong learners *ceteris paribus*" (pp. 84–85). Consistent with our earlier comments in this chapter, Gorard and Selwyn concluded that "the e-learning society seems to be remarkably similar to its non-technological predecessor" (p. 85). "ICT is *increasing* levels of participation within the social groups that were learning anyway" (p. 85, original emphasis).

When considering age as a "digital divider", Hart, Chaparro, and Halcomb (2008) reported that older adults in the United States are the fastest growing group of Internet users. It should be noted, however, that "older" has a wide range of meaning in the literature; anything from 40 to over 70 years of age. In a multi-disciplinary review of computer use by "older" adults, Wagner, Hassanein and Head (2010) examined 151 articles sourced from a variety of disciplines, from which the authors concluded: "The Education discipline has shown surprisingly little interest in this area, especially considering that continued learning is often listed as a way in which computers and the Internet can be of most assistance to older adults" (p. 872). This echoed an earlier comment from Selwyn and Gorard (2004) to the effect that:

> ICTs would appear to be more motivating for informal learning than formal learning – or at least we can conclude that ICTs are proving more appealing to motivated informal learners than to motivated formal learners. How ICTs work in this way for learners, and how ICTs can be encouraged to work in this way for current non-learners, should now form the basis for sustained investigation.
>
> *(pp. 307–308)*

Greenhow, Gibbins, and Menzer (2015) also commented that "little is known about informal learning processes within out-of-school online contexts" (p. 593).

Thus at the time of writing there is a dearth of educational literature on informal learning through new media of the kind we are considering in this book. We can, however, draw some conclusions about favourable elements that facilitate such learning. For example, Rasi and Kilpeläinen (2016) investigated a diverse group of people in Finland, aged between 62 and 86 years. In terms of the context of learning in this book, the users accessed the Internet for information on hobbies, general topics, and social media. Confidence in Internet use was a critical factor, as was the presence of support in the form of friends and family, or a computer club in the village.

It seems to us that there are key motivations that determine whether an adult will choose to use new media (principally the Internet) to seek information related to informal learning, as opposed to social or business motives. In all Internet searching, confidence is key, and confidence comes with skill in using the Internet. As Litt (2013, p. 628) concluded from a review of users' Internet skills:

> While use of the Internet has the potential to remedy existing social and economic inequalities by providing access to a world of information and people, an inability to navigate around the Internet or make educated judgments about Internet use may have the opposite effect, exacerbating such inequalities. Unfortunately, current research indicates those already in more privileged positions are the ones who may be more likely to reap the benefits from the Internet, such as networking and applying for jobs, while avoiding associated losses and risks, like becoming victims of online scams and accidently oversharing private information. The encouraging news though is that Internet skills can be taught and enhanced through intervention, training programs, and education. Thus, scholars must continue to advance this research.

This emphasises the importance of enhancing Internet skills to increase accessibility of information.

Motivation to Search the Internet

From the stories in this book, we can identify clear uses of the Internet and new media for learning science. These uses are: for checking facts, for focused searching, and for exploration about science and technology that often led to discovery. Each will be discussed separately in the following sections.

Checking Facts

The Internet provides a medium for people to check their new or uncertain knowledge against expert information. For example, Tina can identify bees from home:

> *The on-line identification tool can be used from a home computer. Individuals are encouraged to take digital photographs of any unusual species and upload them to their site. This "BeeWatch" tool provides relatively easy questions to help identify the bumblebee in the photo. Other Internet users are encouraged to check identifications and an expert will also look at examples and provide feedback, although the latter can*

take some weeks. Recently a new tool providing a self-testing set of photographs has been provided with immediate feedback.

From time to time, Keith too will check the Internet to identify an unknown plant, although his preferred source of information is to consult a book. Such identification facilities are often the major tool for citizen science projects, as Liz found:

I've also taken part in a couple of citizen science projects. One I loved was classifying galaxies with the Galaxy Explorer project. It was interesting looking at the different types of galaxies, and thinking about how far away they are, and how amazing our universe is.

Focused Searching

In many cases, searching the Internet may be prompted by a direct question arising from information provided by some kind of authority figure or institution. Without such a clear central authority of information – a doctor, a local museum, or an authoritative teacher – many people approach the Internet bewildered by the enormous array of information available. In Penny's case, although she consulted the Internet, it was not especially helpful:

After saying I wouldn't bother to look up the terms used in my scan reports, I did Google the terms online and the "lacuna infarction" seemed to be associated with strokes, though it wasn't clear whether it indicated I'd actually had a stroke... Philip does lots of online research on statins and it all seems to confirm that they have small benefits for anyone and are more likely to harm women, even if they haven't already had nasty side-effects... I persuaded the radiological clinic to give me a copy of the results, but it wasn't informative. The unknown terms I looked up online were explained by other unknown terms, making it all rather unhelpful... The complexity of medical terminology makes it difficult to sort through information and opinions, and how do you know if you are using the right terms to describe your own symptoms? I have had several medical opinions, and I have my own opinions, but don't have any way to test whether any are right or not.

Paul and Liz, on the other hand, found their Internet searches fruitful, but insufficient. Paul explained:

Of course, the Internet did not disappoint. I found new several "new" species of dinosaur, discussed with Emma various theories about life cycles and habitat and was able to come up to speed reasonably quickly, provided of course that I took websites such as Wikipedia as a starting point and not as the last word. But it wasn't quite enough for Emma so I took her to a Natural History Museum.

Liz, also helping her daughter, said

> *my elder daughter chose mercury and sulphuric acid. We did the work, checked and rechecked my notes, researched on the Internet, but something wasn't quite right. Thankfully the last thing our course facilitators had said was to contact them if we needed help.*

Warren is an intermittent user, preferring books to the Internet, but he also consults favoured sites:

> *I also have a current reading list for the guides I train and I use websites and Google when wanting quick updates. Then there are TV documentaries, museums, guide colleagues and, quite often, the guests themselves, as other sources of information.*

Exploration and Discovery

Nearly all of our case story authors talk of the Internet as a crucial vehicle for exploring a topic once it had become important to them. Indeed, even when they were not able to use it, the Internet looms in their writing. Richard, for example, notes how much easier his task would have been if the Internet had been available at the time he was researching the Baird televisor. While the Internet and new media are crucially important for the focused search for facts, evidence and information regarding the problem at hand, it is also important to note that they can play other roles, in particular leading the information-seeker along other paths hitherto unconsidered. Ana, for example, consulted the Internet about her health in pregnancy and learned, on further exploration, about the health of the child:

> *Neonatal diabetes is unlikely but the child can develop other diseases such as obesity, intolerance to glucose, or cardiovascular disease. So I decided to change my diet, reading all the labels in order to choose, for example, the best type of yogurt or bread.*

Liz, too, found further information from an initial site search:

> *looking for information to help with yet another piece of homework [I] was led to a program called "Future learn" which had a free course called "Kitchen Chemistry". It had been put together by the University of East Anglia. It was almost a continuation of the chemistry work we had done in the program.*

Ketan is accustomed to randomly accessing a range of offerings downloaded from the Internet in the spirit of exploration:

long drives were accompanied by endless talks and podcasts exploring religion, science, technology and psychology. This was my scientific training – I cut my teeth on a range of weird, unbelievable pseudosciences.

The people in our stories learned from their exploration, but the pathway had to be relevant to the task at hand. This is entirely consistent with the kinds of learning we have described in this book.

As many of the cases explored in this book have demonstrated, self-directed enquiry often depends on a moment of discovery, a moment when we are surprised about the world and motivated to know more. Importantly, while the Internet and new media can play a significant role in the deeper exploration and testing of knowledge, it can also play a role in such discovery, in laying the groundwork of surprise to be later followed by deeper learning. This was the case for Liz:

Over the years I've been spurred on to keep learning and trying new things. I wanted to keep on being able to balance chemical equations, so I'd do worksheets that I found on the Internet… I found the University of Nottingham's Periodic Tables of Videos with Prof. Sir Martyn Poliakoff and Brady Haran. That led me to Sixty Symbols.

In searching for information about bees, Tina found the Bumblebee Conservation Trust and, at the same time, a like-minded community:

In 2012, the Trust started a national "Bees for Everyone" project to raise public awareness of the threats bumblebees face and to help rare bumblebees through active conservation work to safeguard, restore, and create habitats for them. Part of this was to ask for volunteers to walk a self-chosen transect to identify bees… I was delighted as I could continue to develop my new skills as well as choose an area that interested me. Consequently, I have been doing a "BeeWalk" in Aylestone Meadows since 2013…. However the combination of on-line interaction with some opportunity to meet other people interested in the topic is ideal for me.

Thus the Internet use by our group of narrators was initially generally informed and focused, but often led to new and sometimes unexpected information. They consulted appropriate sites with the general expectation (not always realised) that they would find the information they needed. They had considerable confidence in their ability to do this, which enabled an analytical and critical approach to the websites concerned. This extended use of the Internet, sparked by an initial search for information, is exemplified by this story from Will Grant, consultant for this chapter.

Every Christmas my family (two children then aged four and six years and my wife and I) spend a few weeks visiting both sets of grandparents. Staying at my parents' house, my four-year-old daughter and I were rummaging through an old toy box and

> *found a tiny, key ring sized Rubik's cube. With all sides in the nicely aligned finished position, she was fascinated by this thing – she loved the colours and the shape of the object, and the way that you could turn it in any direction. We played for a bit slightly messing it up (turning one or two sides), and then bringing it back to normal.*
>
> *But then I made a mistake. I twisted too many sides… and the cube was a mess. She looked at me; her eyes widening that little bit… She'd realised it could be broken, and oh so easily. "Don't worry," I said too quickly, "I'll get it back".*
>
> *I went straight to the Internet. Googling "how to complete a Rubik's cube", at first I hoped there would be some sort of simple step by step set of instructions – nothing to learn, just an easy path back so she could play with it. The written instructions I found were obtuse – poor translations of a 3-dimensional puzzle into written text. So I moved over to YouTube (where a search for the string "how to" brings up 448 million results). The first video I found seemed unclear to me, but the next was the first of a series of 7 videos, slowly going through an algorithm you can use to complete the cube from any position. I started in on the first video and the first step, and it didn't seem so hard. So I kept going.*
>
> *And that was the next three days of my holidays, watching each of the seven videos in the sequence, practicing each step. I'd watch again, and practice some more. Then I'd move on to the next step. Then I'd go backwards. (Then for something different, I'd move on to histories of the Cube, and other, faster methods to solve it.)*
>
> *Eventually I solved the tiny key ring sized cube, and got the smile I was hoping for… But I also got something more – by surprise, learning had crept up on me. I had a motivation from the outside, but a lot of what I learnt was made possible by the Internet.*

Will began with a focused search on how to solve a Rubik's Cube. His search led him to YouTube, where he eventually found a series of helpful videos. Then he found himself looking at histories of the Cube and other interesting techniques, which was a progression from focused searching to exploration. His learning progression involved not only watching the expert solvers, but practicing and cross checking as he succeeded in moving to the next step. Like Liz and Paul, who set out to help their children, Will, to his own surprise, found that *"learning had crept up on me"*.

Thus the Internet can be considered a mine ripe for random exploration and discovery, where a series of clicks can lead to a surprising topic and a new area of knowledge. Buntting, Jones, and Cowie (2018) discussed how visitors to the Science Learning Hub (sciencelearn.org.nz) often found themselves involved in what Buntting et al. described as "incidental learning", serendipitous excursions into new areas stimulated by exploration of the information that was intentionally sought.

Are the Internet and new media qualitatively different from more traditional media or social organisations on this front? The argument can be made that compared with traditional media and forms of social organisation – organised so much more around geographically bound rather than interest bound communities – the Internet offers significantly more space for unusual topic discovery. This has clear positives. After all, is it not a social good if people are able to find and pursue their own interests along their own pathways? Yet it also has some more worrying implications, which are discussed later in this chapter.

Learning About Science Through New Media: Social Networks

There is an important element in Internet learning that relates to the more general communication of science through online environments known as social networks. There is a large body of scientists who communicate through YouTube, Facebook, and other social media and who have an enthusiastic following of many millions of viewers and readers. Brossard (2013, p. 14097) stated that "American lay audiences are increasingly using Internet sources outside of mainstream journalistic channels for science-related information". These sources include blogs, videos, and social network comments. There is strong evidence that such communication enhances the profile of the scientists concerned (Liang et al., 2014) but this also carries the risk that "open and interactive dialogues inherent to Web 2.0 tools like Twitter and Facebook enable audiences to repurpose and translate scientists' research findings using their own interpretations and debate them on social media" (p. 24).

Brossard (2013, p. 14099) made the point that "those who tend to pay attention to science content online tend to be more knowledgeable about science, more educated and primarily male". These demographics dictate the *kind* of information sought online, which tends to relate to new technologies, to controversial issues such as nanotechnology or climate change, and perhaps the latest news. There is, however, some evidence that interested participation in Internet information-seeking may include people from less well-educated groups (Cacciatore, Scheufele, & Corley, 2014; Eynon & Helsper, 2011).

The value of such networks in learning about a particular area of science is exemplified by two studies designed to probe science learning through social media. Fauville, Dupont, von Thun, and Lundin (2015) monitored a marine institution's Facebook page, examining the learning opportunities provided by this engagement. They found that, despite a solid support base for the page more generally, participants did not engage very much with the opportunities for discussion, either with the institution or with each other. Rather, they scanned the page for interesting information in a manner comparable to other websites. It seems that the presence of "strangers" on the page discouraged such interactions.

These authors pointed out that, as far back as 1994, Wagner made the following comment: "The growing 'folk acceptance' of a causal relationship between system interactivity and instructional interaction has placed an unreal expectation on interaction technologies to ensure that instructional interactions do occur" (as cited in Fauville et al., 2015, p. 68). Thus there is no guarantee that interactive technology will necessarily result in learning because of the enhanced ability to interact with the user.

It seems reasonable to conclude that the concept of sharing information with like-minded people, so important to Facebook users, has to be actively fostered by such institutional websites. For Kristen's place of employment, this was indeed the case: *"We share a lot of our knowledge and interests – we have a staff science Facebook page started by a young Explainer, where staff post various scientific articles for us to look at."*

We note in passing that where groups of people have a common interest and are not "strangers", institutional online forums are commonly attached to formal online courses.

A study by Greenhow, Gibbins, and Menzer (2015) reinforced the importance of group involvement. They examined use by high school and college students of a Facebook site concerned with climate change. Only a minority engaged in discussion, indicating that while sophisticated engagement with learning outcomes can occur on social media sites, it is not the norm. Of course we have no way of knowing if passive monitoring of these discussions also resulted in meaningful learning. Greenhow et al. added, however: "It may be that such social media applications will be most effective with a core group of highly interested, niche users rather than broadly applicable to all users" (p. 602). There are uncountable niches, but the mere existence of a social media page will not guarantee that the users will form a coherent group. There has to be a connection outside of the medium, and this was the case with Tina, who wrote: *"It is important to take full opportunity of increasing on-line support and chance to meet other interested learners."*

The personal engagement that is part of comments and discussion on social media may further encourage learning about science but, in general, social online participants are not seeking specific information of the type described in our case stories. Searching for such information at a time of need is unlikely to take place on a social media site or to result in the protracted series of discursive interactions mooted by researchers as characterising effective learning through social media networks. Nevertheless, for the casual user there may be snippets of information that are valuable: Tiki, for example, said that she often appreciates the views and experiences of others:

> *Then there's the websites which aren't meant to be objective and authoritative, part of the New Web. User forums, wiki-style maps, social knowledge collators. Sites like "GrowStuff", which aims to collect and collate information on what crops are being grown at what times of year in highly localised areas. You can look at other gardeners close to you and see what kinds of things they're growing, and whether it worked or not. Sites like scrumping maps or "Ripe Near Me", which lets people put up abandoned fruit trees or their own fruit trees on a map as a way of sharing the produce – and coincidentally creates a collection of knowledge about local edibles. Sites like the blog sites of bush survival experts and teachers, who post about interesting plants they encounter as a way of engaging with the community they hope to teach. When I read these, I know that I'm getting anecdata rather than scientific data – but maybe if I want to grow lemons in Melbourne I don't need research, I need to hear from someone*

> *who tried it and got it right. These sites can fail the "10% of your brain" test, in that they can propagate any popular error because people "feel it's right" and "know it's true" so happily repeat it. But they do allow a crowdsourcing of intelligence that – if sieved gently – can turn up all kinds of useful things.*

Selwyn and Gorard (2004) also found that their respondents engaged in this kind of exploration:

> Also evident from our interviews, but more difficult to pin down and maybe under-reported, was the range of 'unintentional' informal learning which was taking place via ICT – learning which was smaller in scale but incremental in its effects… It was clear that for some people, ICTs such as the Internet and computers were contributing in this way to such passive education… Yet… there was little evidence that ICT has created a new-found desire for learning – rather that it was building upon previous learning behaviours and dispositions.
>
> *(pp. 304–305)*

We have discussed research into social media in this chapter noting that further findings will emerge about ways to increase public participation online in scientific matters. This is a dynamic and important area of investigation, likely to provide recommendations for best practice not only for institutional social media sites but for all websites designed to attract and engage the public.

Hazards of New Media

It would be remiss not to note that searching for information on the Internet carries an inherent risk that the information will be wrong. Since perhaps the beginning of the Internet era, a range of thinkers and theorists have offered near Utopian thinking on what the Internet has meant for society. At the extreme, radically free information-sharing offers a way to democratise governments, to soften capitalism, to usher in transformative new societies or indeed epochs (for example, Kurzweil, 2005) – and to let anyone learn anything they want to learn.

Yet while this promise is by no means ruled out, it is plainly apparent that the last few years have proved a very bumpy road to Utopia. We have seen societies not brought together in a social media landscape, but increasingly polarised; we have seen movements of like-minded iconoclasts (to use a rather generous term) reinforce and trumpet each other's unscientific descriptions for key social and political issues. To be frank, for the unwary, learning on the Internet can be a dangerous thing. Not only are the roads poorly signposted, but roads that lead away from accepted science abound.

Ketan has used social media professionally, to try to understand what drives those who oppose scientific advances:

> *My day job involves the digestion and dissemination of scientific information. Sometimes, my prolific social media and blogging commentary serves a therapeutic function – a sarcastic appeasement for those of us growing weary of the lack of science in political leadership. But when it comes to finding ethical, analytically honest solutions to the roadblocks facing clean technology, the solutions lie in understanding the motivations of those who react to widespread technological change. These are the tangible consequences of a more compassionate approach to scepticism, tempered, of course, by a willingness to always call out falsehoods, where it's appropriate and useful.*

Brossard (2013) stated that "empirical evidence seems to suggest that inaccurate accounts of scientific phenomena are present in the online world although exact proportions have yet to be quantified across the multitude of online platforms" (p. 14099). Despite this, she is optimistic about Internet use. "Motivated individuals are not passive when selecting which science stories to pay attention to online" and "when motivated to learn about a specific scientific issue, these information-seeking individuals tend to prefer messages presenting two sides of an argument" (p. 14099).

Does every learner have the capacity to evaluate the worth of what is offered online, as argued for in constructivist thinking? Here a constructivist approach suggests that acquired schemas of bad learning will reinforce one another as readily as good learning. In a seminal paper, Metzger and Flanagin (2013) concluded that "one consequence of the increased information abundance is the accompanying issue of finding the best information to meet one's needs from among the enhanced number of possible information providers" (pp. 211–212). These authors listed a range of ways that the information provided may not be credible, including lack of filtering through professional gatekeepers; being out of date or incomplete or inaccurate; and lacking attribution to a reliable source. Further, they noted, "there are few standards for quality control and evaluation online" (p. 212). These authors commented that:

> A common assumption about how people evaluate online information is that people are motivated to evaluate the information they receive to determine its trustworthiness. A corollary of this assumption is that people engage in effortful evaluative processes in order to be certain of information or source credibility.
>
> *(p. 213)*

Metzger and Flanagin (2013) suggested that factors such as the accuracy, authority, objectivity, currency, and coverage of websites were important indicators of its credibility, "Yet, research shows that people rarely engage in effortful information evaluation tasks, opting instead to base decisions on factors like web site design and navigability" (p. 213). Tiki is aware of these pitfalls:

> *Sure, there's a lot of dross. Especially if you want to know about the uses of plants – "natural medicine" misinformation abounds. My general science training kicks in here nicely, and I try*

> *to be thorough about vetting my sources, cross-checking information, and making sure the sources I'm cross-checking against aren't just plagiarising each other. I follow the rule of thumb of "Never trust Wikipedia until you've read the research". Also the rules "Pliny and Aristotle didn't always know what they were talking about", "just because someone gave two plants a similar common name doesn't mean you can conflate their chemical properties" and "if it sounds like a great story, someone might have made it up".*

To judge the value of an Internet site requires drawing conclusions about its validity. Internet users must reduce the inevitable overload of information and sort through the *"dross"* to find relevant and useful material.

Judging a Credible Source

To reduce information overload, Metzger and Flanagin (2013) stated that various theories related to information processing "suggest that Internet information consumers are likely to cope with the perceived costs of information search and overload by using strategies that minimise their cognitive effort and time, through the use of cognitive heuristics" (p. 214). These authors listed a number of such heuristics or "rules of thumb" commonly employed. These "short cuts" to learning, informed by experience, are:

- The reputation heuristic: name recognition of the source. The more authoritative the source the more it is trusted, provided the information-seeker is familiar with the name.
- The endorsement heuristic: trusting a site which is known to be trusted or recommended by others.
- The consistency heuristic: checking across various sites to see if the information is in agreement. This is sometimes termed the "consensus heuristic".
- The self-confirmation heuristic: trusting information if it confirms pre-existing beliefs, attitudes, or opinions.
- The expectancy violation heuristic: if the website does not meet expectations, the information is not credible.
- The persuasive intent heuristic: belief that the information is overtly designed to persuade. This often applies to commercial sites.

(pp. 215–216)

The degree to which the user will apply these various "rules of thumb" depends to some extent on the information being sought. Metzger and Flanagin (2013) point out that when seeking information about, say, health matters, the credibility of the source may undergo much more careful evaluation than information related to entertainment. Considering these six ways of judging a website, the first three are often cited as valid ways to evaluate the information provided. This is discussed in more detail in the next section. The endorsement heuristic, can however, sometimes be problematic. It can lead to group approval when the site is, in fact, presenting wrong science – just because others support a site does not make it reliable. This is closely related to the self-confirmation heuristic, which is powerfully influenced by people's own beliefs in the science.

The theory of cultural cognition has important contributions to offer in this regard. According to Kahan (2011) when talking about climate science, this theory "predicts that individuals will more readily recall instances of experts taking the position that is consistent with their cultural predisposition than ones taking positions inconsistent with it." (p. 149). Kahan added that "individuals tend to search out information congenial to their cultural predispositions" and that they might well work harder to locate such information:

> As such, scientific consensus cannot be expected to counteract the polarizing effects of cultural cognition because apprehension of it will necessarily occur through the same social psychological mechanisms that shape individuals' perceptions of every other manner [*sic*] of fact.
>
> *(p. 150)*

This aspect of the information-seeker is entirely consistent with constructivism and with science communication theories. All information is filtered through the personal framework of the individual. Every time someone accesses the Internet, judgements have to be made, and it is evident from our case stories that such judgements were indeed made by those who used the Internet as a source of information. Amongst our stories we can find a clear thread of adults using critical thinking skills developed in other parts of life to approach sources on the Internet.

Helping People to Learn From the Internet

This leads us to the question of what, apart from confidence, might assist in gaining knowledge from the Internet? Eynon and Helsper (2011) found that when considering fact checking, "Internet users who have higher levels of Internet self-efficacy, have more positive attitudes towards ICTs, are better educated, and are younger, are more likely to use the Internet for fact checking" (p. 542). Similar characteristics apply to online informal learning, with the additional inclusion of those who have access to the Internet at home.

Thus confidence, levels of education, and age were significant factors in learning from the Internet although levels of education proved less useful as a predictor of learning than expected. Having Internet access at home, however, proved very important. Nevertheless, these authors stated that: "How and why people choose to use the Internet for learning instead of or to complement other ways of learning remains unclear" (Eynon & Helsper, 2011, p. 547). They added that "supporting online learning opportunities is not straightforward" (p. 548).

Selwyn and Gorard (2004) found, overall, that participation in ICT-based learning does not depend so much on matters such as accessibility or time, but on "the fundamental issues of motivation and disposition" (p. 307).

With the case stories of this book in mind, it is clear that organisations that wish to support motivated adults to learn about science and technology can use the Internet to assist in a range of ways. Important skills for the learner include critical thinking and search literacy and these skills must be fostered to encourage successful interactions with the Internet. Critical thinking skills include the ability to assess websites for credibility of their sources and there are many lists of such assessments in the literature (see, for example, Knowles, Holton & Swanson, 2015, pp. 214–215); and many university sites, for example, The University of Edinburgh (n.d.). Such lists include assessing the audience for the website, its authority and credibility, its overall presentation, and its objectivity.

A very detailed analysis of evaluation criteria methods is given by Hasan and Abeulrub (2011) who reviewed and synthesised over 50 papers on this subject published over the previous two decades. They described four broad categories of analysis: content, design, organization, and user-friendliness. All assessment criteria have elements that closely relate to the heuristics people use to judge websites listed by Metzger and Flanagin (2013), and it would be desirable if information-seekers were aware of their own innate biases in making such judgements.

Beyond the ability to judge these aspects of a website, however, are the inherent cultural biases described by Kahan (2011). "To overcome this effect, communicators must attend to the cultural meaning as well as the scientific content of information" (p. 169). Finding ways to present information in such a manner that it is consistent with personal and cultural considerations is more likely to have a positive outcome. The values held by experts are also problematic if they appear to conflict with personal values. Finally, "narrative framing" (pp. 169–70) in such a way as to be culturally congenial is imperative for acceptance of science. All these are basic tenets of science communication and anyone wishing to assist others in finding information from the Internet will need to understand the importance of listening to the sub-text of an individual's quest for knowledge.

The skill of "search literacy" is stated by Scott and O'Sullivan (2005) to mean not only the ability to navigate, but a deeper understanding of structure and purpose:

> We have observed students get frustrated and end up surfing from one site to another or abandoning their search in frustration. Understanding the nature of information, the structure of the Internet, and the ability to efficiently and effectively navigate the hypertext environment of the Internet are critical information skills that teachers need to teach and high school students need to master.
>
> *(p. 21)*

A simple keyword search is unlikely to yield the best and most useful information and may lead to abandonment. For all ages of users, interpreting navigational symbols may be problematic and it takes experience to search successfully:

> The interpretation of the semiotic symbols on the Internet, including graphics, text, sound, icons, images and video, requires a specific knowledge structure and may pose difficulties for certain types of learners … many high school students do not possess an understanding of how to develop a search strategy or how to refine a search beyond using keywords.
>
> *(pp. 22–23)*

In a comprehensive review of the literature on Internet skills, Litt (2013) found that generally, research has focused either on the ability to perform specific tasks (send an email, download music…) or on self-efficacy measures. Self-efficacy is discussed in more detail in the final chapters of this book. Suffice to say here that perceptions of self-efficacy varied widely across groups but there was a gender effect in this domain. An interesting finding from several studies was that, although younger people might have more technical skills, older adults tended to perform better on information and evaluative skills (Litt, 2013, p. 621). Research in this area needs to recognise and identify how skills vary among different people. "From a scholarly perspective, a more nuanced approach will help identify how specific Internet skills relate to other factors (e.g. online participation)" (p. 625). Litt concluded that:

> while use of the Internet has the potential to remedy existing social and economic inequalities by providing access to a world of information and people, an inability to navigate around the Internet or make educated judgments about Internet use may have the opposite effect, exacerbating such inequalities.
>
> *(pp. 625–626)*

In general, these studies did not look at science learning, but we may conclude that it is important that people seeking information from the Internet are aware of how to search effectively. This skill is clearly critical. Further, they need to have confidence in their ability to achieve a satisfactory outcome.

To these fundamental skills we would add the more subtle aspects recommended by Selwyn and Gorard (2004, p. 308) – it is important to understand how ICTs "complement or usurp" more traditional sources such as books, television and social networks, and how the increased mobility of such sources facilitates learning in the home and in leisure time. We also need to understand how the outcomes of such learning differ from more traditional methods of education. We will discuss these issues further in the final chapters of this book.

Acknowledgement

We thank Dr Will Grant of The Australian National University for his insightful contributions to this chapter.

References

Australian Bureau of Statistics. (2016). 8146.0 Household Use of Internet Technology, Australia, 2014–5. Retrieved from http://www.abs.gov.au/ausstats/abs@.nsf/mf/8146.0

Brossard, D. (2013). New media landscapes and the science information consumer. *Proceedings of the National Academy of Sciences*, 110, 14096–14101.

Buntting, C., Jones, A., & Cowie, B. (2018). Using a digital platform to mediate intentional and incidental science learning. In D. Corrigan, C. Buntting, & A. Jones (Eds.), *Navigating the changing landscape of formal and informal science learning opportunities* (pp. 171–182). Dordrecht, The Netherlands: SpringerNature.

Cacciatore, M.A., Scheufele, D.A., & Corley, E.A. (2014). Another (methodological) look at knowledge gaps and the Internet's potential for closing them. *Public Understanding of Science*, 23, 376–394.

Cook, J., & Light, A. (2006). New patterns of power and participation? Designing ICT for informal and community learning. *E-Learning and Digital Media*, 3, 51–61.

Eynon, R., & Helsper, E. (2011). Adults learning online: Digital choice and/or digital exclusion? *New Media and Society*, 13, 534–551.

Fauville, G., Dupont, S., von Thun, S., & Lundin, J. (2015). Can Facebook be used to increase scientific literacy? A case study of the Monterey Bay Aquarium Research Institute Facebook page and ocean literacy. *Computers and Education*, 82, 60–73.

Goody, J. (1977). *The domestication of the savage mind*. Cambridge, UK: Cambridge University Press.

Gorard, S., & Selwyn, N. (2005). Towards a le@rning society? The impact of technology on patterns of participation in lifelong learning. *British Journal of Sociology of Education*, 26, 71–89.

Greenhow, C., Gibbins, T., & Menzer, M. (2015). Re-thinking scientific literacy out-of-school: Arguing science issues in a niche Facebook application. *Computers in Human Behavior*, 53, 593–604.

Hart, T. A., Chaparro, B. S., & Halcomb, C. G. (2008). Evaluating websites for older adults: adherence to 'senior-friendly' guidelines and end-user performance. *Behavior and Information Technology*, 27, 191–199.

Hasan, L., & Abuelreb, E. (2011). Assessing the quality of web sites. *Applied Computing and Informatics*, 9, 11–29.

International Telecommunication Union. (2015). Key ICT indicators for developed and developing countries and the world (totals and penetration rates). Retrieved from http://www.itu.int/en/ITU-D/Statistics/Documents/statistics/2016/Individuals_Internet_2000-2015.xls

International Telecommunication Union. (2016). *ICT Facts and Figures 2016*. Retrieved from http://www.itu.int/en/ITU-D/Statistics/Documents/facts/ICTFactsFigures2016.pdf

Kahan, D. M. (2011). Cultural cognition of scientific consensus. *Journal of Risk Research*, 14, 147–174.

Knowles, M. S., Holton, E. F., III, & Swanson, R. A. (2015). *The adult learner* (8th ed.). New York: Routledge.

Kurzweil, R. (2005). *The singularity is near: When humans transcend biology*. London: Penguin.

Liang, X., Yi-Fan Su, L., Yeo, S. K., Scheufele, D. A., Brossard, D., Xenos, M., … Corley, E. A. (2014). Communicating science in new media environments. *Journalism and Mass Communication Quarterly*, 91, 772–791.

Litt, E. (2013). Measuring users' Internet skills: A review of past assessments and a look toward the future. *New Media and Society*, 15, 612–630.

Metzger, M. J., & Flanagin, A. J. (2013). Credibility and trust of information in online environments: The use of cognitive heuristics. *Journal of Pragmatics*, 59, Part B, 210–220.

Office of National Statistics UK. 2015. *Internet users: 2015*. Retrieved from file:///Users/willgrant/Downloads/Internet%20users%202015.pdf

Rasi, P., & Kilpeläinen, A. (2016). Older people's use and learning of new media: A case study on remote rural villages in Finnish Lapland. In J. Zhou & G. Salvendy (Eds.), *Human aspects of IT for the aged population. Healthy and active aging*. ITAP 2016. Lecture Notes in Computer Science, vol. 9755. Cham, Switzerland: Springer.

Scott, T. J., & O'Sullivan, M. K. (2005). Analyzing student search strategies: Making a case for integrating information literacy skills into the curriculum. *Teacher Librarian*, 33(1), 21–25.

Selwyn, N., & Gorard, S. (2004). Exploring the role of ICT in facilitating adult informal learning. *Education, Communication and Information*, 4, 293–310. Shirky, C. (2008). *Here comes everybody: The power of organizing without organizations*. London: Penguin.

Smith, A. (2017). Record shares of Americans now own smartphones, have home broadband. Pew Research Center. Retrieved from http://www.pewresearch.org/fact-tank/2017/01/12/evolution-of-technology/

The University of Edinburgh. (n.d.). How to evaluate website content. Retrieved from http://www.ed.ac.uk/information-services/library-museum-gallery/finding-resources/library-databases/databases-overview/evaluating-websites

Wagner, N., Hassanein, K., & Head, M. (2010). Computer use by older adults: A multidisciplinary review. *Computers in Human Behavior*, 26, 870–882.

10

SUPPORTING SELF-DIRECTED LEARNING IN SCIENCE AND TECHNOLOGY

This chapter synthesises the preceding chapters to focus on how self-directed adults can be supported in their learning. Several major models of learning were outlined in Chapter 2 so that we could identify, in a theoretical sense, the challenges that would be faced by adults who needed to learn more science and technology to meet a personal need. The empirical core of this book is the series of case stories of adults' self-directed learning, recorded in Chapters 3 to 6, that explored how these learners pursued an understanding of science-related issues that arose, sometimes unexpectedly, in their lives. It became very clear that all were fully in charge of their learning. This fact underpins this chapter and indeed this book as a whole. We are focusing on self-directed learners, adults who have a personal need to learn, who know what they want to know about, albeit they sometimes may not be sure about where to begin, where to go next, or how to focus their search. But they did not all learn in the same way, and the Moon Diary assignment described in Chapter 7 illustrated the enormous diversity in how adults approach a learning task when they are free to tackle it in any way they choose. Chapters 8 and 9 were devoted to reviewing how our case story authors used various resources as they worked towards achieving their learning goals, and identifying the factors that were helpful and those that were not. Our authors chose the resources they tried and even when given advice from a friend or mentor, they still could choose whether or not to take that advice. They searched for, and they picked and chose among, potential resources because they were interested to learn about their chosen topic and motivated to solve the issue that was central to their need to know. It was their need to know, underpinned by their motivation and confidence and self-belief, that enabled them to embark on a journey to resolve, or at least come to terms with, their need.

DOI: 10.4324/9780429434303-10

In this chapter we pull together all of these findings to identify how other people – educators, guides, counsellors, or friends and colleagues – can best help adult learners to find and use the resources necessary to resolve their learning need. First, we revisit the list of learning challenges we identified in Chapter 2 and test the validity of this list against the findings of Chapters 8 and 9. This enables us to emphasise four essential skills that adults need to become effective self-directed learners. Second, we explore further the personal resources that allow learners to be persistent and confident that they can achieve their goal by revisiting and discussing the significance of the learners' motivation – first mentioned in Chapter 2 – the nature of their engagement, and their need for positive self-efficacy. Third, we review our case stories to identify and describe the kinds of learning relationships that assisted them to resolve their issue. Then, in the final part of the chapter, we set out the ways through which other people can support self-directed learners to take the actions that can make their learning task easier, thereby ensuring that, as far as possible, learners persist in their self-directed study to achieve their goals. In the next and final chapter, we review and reflect upon our findings to suggest how school curriculum might change to ensure that students develop the self-directed learning skills they will need as adults to cope in a changing world.

Essential Skills for Effective Self-Directed Learning

In Chapter 2 we synthesised the common threads in three learning models – constructivism, andragogy, and heutagogy – to identify the main challenges that an adult learner would face in tackling a chosen learning task. These challenges required learners to do the following:

- Identify the specific knowledge and skills that form part of the chosen sector of science and technology.
- Develop personal criteria for success in respect of the knowledge and skills to be acquired and in respect of the particular reasons for acquiring it.
- Draw on their prior experiences that seem relevant to the chosen task.
- Make direct and immediate use of the knowledge they acquire to solve specific problems.
- Reflect on what was involved in gaining the desired knowledge, and deciding "Have I learned enough, at least for the time being?"

How well does this list of challenges match the activities that our case story authors actually undertook to solve their learning needs? We found this list lacking in one essential feature: finding and using the resources that will provide the information learners need. The reason for this lack is obvious: The list was drawn from models of learning that were developed essentially in a teaching context, where the resources are either provided by the teacher or the curriculum, or the learner is directed to them. Self-directed learners who are not participating in a course but have a personal learning need must find their own resources and figure out how to use them. We found that our case story authors recognised and demonstrated the kinds of learning skills that helped them to access and make use of the resources they accessed, even if they were not always successful. These skills were discussed in the previous two chapters, particularly in the Personal Resources section of Chapter 8, and because they underpin the subsequent sections of this chapter, we restate them explicitly here. We suggest there are four clusters of skills learners require to successfully pursue their need to learn and they are described briefly below. These clusters have similarities with the list of challenges above, but place more emphasis on information and resources, and it is in this area that those people who wish to support self-directed learners have a role to play.

Decide on the information needed. Early in their journey learners may not be clear about what information they actually need, so the ability to ask questions, of themselves or others, such as experts in the field, and to persist in finding that information, is a fundamental skill.

Access the information. Whether or not learners can use the information that they acquire depends on its "match" to the learner's background and experience. There needs to be a cognitive match, so the learner understands the information and can make best use of it, and a skills match that can lead to an advance in the learner's current skill level. In addition, and because often the learning task has an emotional dimension, there should be an emotional match – the information should be affectively appealing as learners are unlikely to pursue information that is at odds with their attitudes and values.

Make the best use of information. There are two aspects to making best use of information. The first is to establish its validity. Metzger and Flanagin's (2013) heuristics for judging credible sources are useful here. Second, the information gleaned often needs organising; sifting through what has been found to identify what is useful and credible, then structuring it in a meaningful way. This is a fundamental research skill that needs practice.

Reflect on what is being learned. Reflecting on what is being learned enables learners to determine whether they are making progress towards their goals. It also helps them to understand their own ways of learning so their next step can be more effective.

It is important to point out that although there is a sense of sequence in this list of skill clusters, in practice it is iterative. Learners frequently move back and forward through these steps as they continuously reflect on and assess their progress. Further, this list makes clear that the need for resources to be accessible is fundamental to successful learning. Some directions about making static resources more accessible to learners were provided in the previous two chapters. Attention was drawn to the helpful role another person, perhaps a friend or mentor, could play by increasing access to resources, and we devote attention to these supporting roles later. First we explore the less tangible "personal resources" prerequisite for self-directed learning that was mentioned in Chapter 8: the learners' need for motivation, confidence, and self-belief. We note in passing that, although our focus here is on the learning of science and technology, these personal resources and the essential learning skills described above, are generalisable to any kind of self-directed learning.

Prerequisite Personal Resources for Self-Directed Learners

At a personal level, successful learning will require learners to have positive motivation towards the task that has been chosen, active engagement in the process of learning, and a positive belief in their self-efficacy as a learner. These three attributes determine the individual's capacity for suitable self-direction in respect of any aspirational learning goals. These determinants to the act of learning are inter-related, but here they are presented in the sequence in which they are often initially triggered by the learner. This sequence will also allow us to suggest how they can be acquired and deployed. We begin with motivation.

Motivation Towards the Chosen Task

The literature on motivation is almost completely concerned with the formal schooling system and usually couched in terms of extrinsic motivation and intrinsic motivation, both of which result in some kind of engagement. Knowledge gained under conditions of external motivation is often retained only in the short term, for example "learning just to pass the exam". This generally negative experience has, in part, caused those who are concerned with adult learning to postulate that the most effective form of learning is driven by intrinsic motivation – that originating within the student – if it is to be successful and retained in the longer term. But is there really a case for making the distinction between intrinsic and extrinsic motivation when considering adult learning?

As explained in Chapter 2, Ryan and Deci (2000) distinguished between kinds of motivation as follows: Intrinsic motivation is that which motivates people to pursue knowledge freely with no sense of external reward, while extrinsic motivation carries connotations of external positive outcomes. For each of our case story authors, the initial motivation to pursue knowledge about a specific scientific topic came from within. There was no extrinsic demand for learning made upon the learners. Nevertheless, there was often an external stimulus: either another person needed help, or there was a level of knowledge required for new employment, or a community issue arose that needed greater understanding. In the cases of the hobbyists, for example, although there was no external factor driving their pursuit of knowledge, there was high interest in achieving the outcomes and those outcomes were personally important and rewarding. It seems to us that all our learners were intrinsically motivated to follow a learning path, whether – like our case story authors – it was a self-imposed one, or – like the Moon Diary students – a path with external goals. Their choice was to learn about science in some highly personal and relevant way. So we do not see the clear divisions between intrinsic and extrinsic motivations as being useful in our analysis of how our adults set about learning science. The important prerequisite was that they were motivated to learn: They knew their ultimate goal and they were actively engaged in their learning.

Active Engagement in Learning

McKinnon and Vos (2015) argued that engagement is "a necessary condition or precursor to participation [in learning]" (p. 300). There is said to be three dimensions to engagement: behavioural, affective, and cognitive. According to Fredricks, Blumenfeld, and Paris (2004),

> behavioral engagement draws on the idea of participation .... Emotional engagement ... is presumed to ... influence willingness to do the work. Finally, cognitive engagement draws on the idea of investment; it incorporates thoughtfulness and willingness to exert the effort necessary to comprehend complex ideas and master difficult skills.
>
> *(p. 60)*

Learning about science and technology would seem to fit into the cognitive dimension, but the other two dimensions also had roles to play in the case stories in this book. There were aspects of affective or emotional involvement in our authors' learning, as well as aspects of action and behavioural change. Fredricks et al. (2004) made the point that these aspects overlap: Engagement can then best be viewed as a "metaconstruct" (p. 61) linking these three dimensions.

The multidimensional aspects of engagement have not been thoroughly explored. For example, Illeris (2002) alternatively represented three dimensions of learning as cognition, emotion, and environment (where environment represents the social aspects of learning), and suggested that all three are always present in successful learning activities. In contrast, McKinnon and Vos (2015) represented engagement as a unified "threshold concept", where this means being "transformative, irreversible, integrative, bounded and likely to involve 'troublesome knowledge'" (Meyer & Land, 2003, as cited in McKinnon & Vos, 2015, p. 307).

In overview, McKinnon and Vos (2015, pp. 308–310) argued that the transformative nature of engagement is tightly connected to self-efficacy. It conveys a new way of seeing the problem and one's relationship to that problem, as demonstrated in the case stories in this book. Engagement is irreversible; it fosters further information-seeking and, in many cases, active involvement. It is integrative, in that it enables connections to be made between ideas, and wider understanding to be achieved. It is bounded; the idea that engagement has frontiers that must be crossed and boundaries that guide the exploration is also further discussed below. Last, it involves "troublesome knowledge". In every case in this book, the participants not only wanted to know, to find out about their troublesome issue, but they found the process challenging.

The idea of engagement as a concept may seem unusual, but the dimensions outlined here do indeed apply to our participants. To describe engagement in these terms may thus be more fruitful than trying to separate it into components which inevitably interact and overlap. We are, therefore, now in a position to say that adult learning requires motivation and engagement, however one might choose to define them. The next requirement is that the learner must have self-belief in their ability to find and use the knowledge – their self-efficacy must be high.

Self-Efficacy as a Learner

The confidence that one can learn something stems from past experience. Echoing Dewey, Jarvis (1987, p. 164) stated that all learning commences with experience. He was here referring to the phenomenon now known as experiential learning but his model of learning begins with "self with biographical history" (p. 166). The translation of this idea of initial experiential relevance into constructivist theory is a natural extension for theories of formal learning. In our case stories, it was knowing how to learn that was the most influential part of the learners' prior experience. Our authors knew that, if they persisted, they should be able to understand and apply what they learned. In some cases they asked for expert help, in others they persevered on their own, but everyone had a strong self-belief that they could conquer the subject matter that was required. Without that belief, we do not think they could have embarked on their individual learning journeys, and they certainly would not have persisted.

Bandura (1977) conducted seminal research into the idea of self-efficacy. He related perceived self-efficacy to a person's own judgement of their capability to plan their actions to attain designated goals. He stated that "expectations of personal mastery affect both initiation and persistence of coping behaviour. The strength of people's convictions in their own effectiveness is likely to affect whether they will even try to cope with given situations" (p. 193). Further, "efficacy expectations determine how much effort people will expend and how long they will persist in the face of obstacles and aversive experiences. The stronger the perceived self-efficacy, the more active the efforts" (p. 194). If attempts to learn are rewarded with failure, however, it then becomes even more important that the learner persists because success effectively raises a person's expectations of mastery, while repeated failure lowers them. Early failures are particularly problematic. "The effects of failure on personal efficacy therefore partly depend on the timing and the total pattern of experiences in which the failures occur" (Bandura, 1977, p. 195).

Bandura made the strong point, however, that encouraging unwarranted self-efficacy belief that cannot be supported by subsequent achievement is very damaging. Similarly, Zimmerman (2000) declared that "perceptions of efficacy depend on a mastery criterion of performance rather than on normative or other criteria .... Self-efficacy beliefs differ conceptually and psychometrically from closely related constructs, such as outcome expectations, self-concept, and perceived control" (p. 84). Perceived control is another construct closely associated to self-efficacy. It "refers to general expectancies about whether outcomes are controlled by one's behaviour or by external forces" (p. 85). This leads us back, in some respects, to extrinsic motivation. Zimmerman (2000) stated that "self-efficacy beliefs also provide students with a sense of agency to motivate their learning through use of such self-regulatory processes as goal setting, self-monitoring, self-evaluation, and strategy use" (p. 87). Positive self-efficacy is the outcome of the successful use of self-regulatory processes.

In every story recorded in this book, the notion of goals for learning was clearly articulated. Every person knew what they wanted to know, even when the way to achieve that knowledge was not clear. Further, our learners monitored their progress; they knew when they needed to change direction or to find new resources and they could self-manage their progress. According to Deci, Vallerand, Pelletier, and Ryan (1991), self-determination (or self-management) is related to control; it implies that a person believes they have choices related to their learning. Merriam, Caffarella, and Baumgartner (2012) agreed that control was very important; self-management "involves learners taking control of and shaping the contextual conditions so that they can reach their stated goals and objectives" (p. 113). These authors described several models of self-directed learning. In the main, these apply to the formal teaching of adults, especially dealing with "learning projects", but some are relevant to the activities described in this book. This is particularly true where the learning is long term and centred on an interest such as a hobby. Merriam et al. cited Berger, who found that her subjects "constantly redefined their projects, changed course, and followed new paths of interest as they proceeded". In essence, the majority of Berger's respondents adopted a trial and error approach, with an emphasis on hands-on experience and practice, guiding themselves both by their successes and their mistakes as they moved on to new levels of learning (as cited in Merriam et al., 2012, p. 113).

What does it mean to give adult learners a free choice to operate, through self-determination, within a given topic in order to follow their own pathway to successful learning? According to Deci et al. (1991):

> Promoting greater self-determination, that is, a greater sense of choice, more self-initiation of behaviour, and greater personal responsibility, is an important developmental goal .... It is becoming increasingly clear that promoting self-determination is the avenue to attaining outcomes such as creativity ... cognitive flexibility ... and self-esteem.
>
> *(p. 342)*

We would hope that, by the time they reach adulthood, individuals would be fully active in the self-determination of their learning. Alas, many school science and technology curricula are still far too overloaded with facts that have to be learned (then most likely forgotten), thus removing the locus of control from the learner – a theme we take up in Chapter 11. Consequently many adults will not have fully acquired the capacity to be self-directed learners, a capacity that can only be acquired by exercising it. The trigger for this will rest in the adult's strength of purpose in acquiring knowledge to resolve an issue.

For many researchers, including Deci and Ryan (1985), self-determination is tightly bound to intrinsic motivation and hence to "enjoyment". For our participants, however, enjoyment was not always evidently a factor in their learning. We consider, therefore, that the notion of enjoyment as a reward may be misleading. "Satisfaction" may be a better term for many adult learners of science. What stands out clearly from all the research on self-determination is that choice is the most important issue. When adults can determine their own learning pathway towards their own goals, they are likely to be more persistent and motivated, and more likely to achieve satisfaction as an outcome.

Thus we can characterise successful adult learners of science as being motivated and engaged: they have high self-efficacy and self-determination. Even while self-determination of learning may be exercised by an individual, it may be directed towards problems that cannot readily be addressed or lead to the use of inappropriate information. Essential steps in the learning process concern appropriate resources and these have been discussed in the previous chapters of this book. Suffice to reiterate here that our learners chose the resources they thought they needed, even when unsure if those resources would actually answer their need. They had complete control over their choice, and if it proved unsatisfactory, they changed their direction and moved on.

Solomon (2003) pointed out that autonomy and independence are very important to autodidacts; they need to have locus of control over their learning. These are the same characteristics that we have identified as significant in our group of self-directed adult learners. Does this mean that we should call them autodidacts? As noted in Chapter 2, Solomon's case studies of autodidacts suggested a resistance to being taught, but our case story learners did not resist being taught. Even though most of them did not take formal courses, they did not hesitate to seek help from tutors of one kind or another when they thought it might help their learning. Importantly, Solomon's series of case studies foreground other features of learning, including that learning is not simply an act of cognition; rather, emotion and cognition act together. "To learn well, we all need a better balance of cognitive, sensory and emotional inputs" (p. 208). The close relationship between attitudes, such as curiosity, and learning is also emphasised by Morris (2015), and both he and Solomon have produced volumes that remind us that all learners are different and need to be supported in different ways, according to their perceived needs.

Partnerships for Learning

Making progress in addressing an issue or solving a problem very often involved our learners consulting someone who provided help in some way in respect of the chosen subject. We pointed out in Chapter 8 that all of our case story authors found it useful to consult other people, sometimes experts in the field, occasionally a teacher, but often friends or other colleagues; in essence, someone who acted as a role model, a mentor, or simply a knowledgeable or empathetic other. In this next section, we explore these kinds of learning relationships or partnerships.

The two best known partnerships for learning are those existing between teachers and their students and between parents and their children, relationships that are, fundamentally, between older, more experienced persons and younger, less experienced persons. With guidance and support from the older person, the younger one learns to become competent in various ways, acquiring cognitive, affective, and psychomotor skills, as well as the interpersonal skills to successfully negotiate their social environment. For most of us, learning relationships with parental figures and various teachers will have been important, often unforgettable, influences on the adults we have become.

Mentoring Relationships

Many learning relationships have a mentoring component and Murray's (2001) straightforward definition provides a good start to discussing the mentoring relationship. "Mentoring is a *deliberate pairing* of a more skilled or more experienced person with a less skilled or less experienced one, with the *mutually agreed goal* of having the less skilled person grow and develop specific competencies" (p. xiii, emphasis added). Although this simple definition comes from the world of business, two properties are both important and generalisable to other areas, including the learning of science and technology. First, the mentorship is deliberate, that is, the mentor and the mentee are both party to the arrangement, and second, there is agreement about the purpose of the relationship: the mentor is to assist the mentee to achieve a mutually agreed goal. These two properties distinguish mentorship from role-modelling, a term that is sometimes confused with mentoring. A mentor and mentee have a two-way relationship; they know each other and, as the mentor guides and supports the mentee, the relationship is usually mutually beneficial. In contrast, the relationship in role-modelling is one way; a person looks up to a role model and tries to emulate him or her but the role model may not necessarily know this person or be aware of the role being played. Of course, role models may become mentors if the people involved know each other and the role model is willing to mentor his or her admirer.

Most authors, including Murray (2001), refer to mentoring as originating in the master–apprentice relationship, so it is not surprising that the research literature infers that two groups – young people and adults in a working environment – are the focus of much of the mentorship that is undertaken. For young people who are not in the work environment, most mentoring programs are aimed at assisting their social and personal wellbeing. Goldner and Mayseless (2008) discussed the similarities and differences between the roles of mentors and other people who care for young people, including parents, therapists, friends, and teachers. There are relationship differences in emotional intensity, usually parent–child relationships are the most intensive and teacher–student relationships the least intensive. Goldner and Mayseless examined these relationships from the theoretical perspectives of attachment, social support, and social learning and pointed out that, at times, the mentoring role can be like that of a parent, and/or therapist, and/or friend, and/or teacher, thus emphasising the need for flexibility on the part of a mentor. The successful mentor must be able to determine the mentee's needs, be they emotional, social, or a lack of knowledge or skills, and respond appropriately to meet those needs. These skills are generalisable to all mentoring relationships.

While we recognise and endorse the mentoring of young people to engage effectively in their social world and, hopefully, become self-directed learners, we focus here on learning relationships involving adults, where formal mentoring arrangements are frequently work and career-oriented. Indeed, most accounts of mentoring relate to organisational arrangements and most of the advice in the literature is proffered to those who might set up such a working relationship. These kinds of mentoring partnerships are frequently found in industry (for example, Allen, Eby, Poteet, Lentz, & Lima, 2004), in teacher education (for example, Desimone, Hochberg, Porter, Polikoff, Schwartz, & Johnson, 2014), in higher education (for example, Yun, Baldi, & Sorcinelli, 2016), and in the health professions (for example, Eller, Lev, & Feurer, 2014).

According to Allen et al. (2004), who conducted a meta-analysis of the literature on mentoring in organisational settings, mentorship facilitates information exchange and knowledge acquisition, including providing access to social networks (p. 129). The mentor was not only a knowledge provider in a career-related sense but an esteemed colleague, providing psychosocial mentoring, such as role-modelling, counselling, and friendship that "centered on enhancing protégé self-esteem, confidence, and identity" (p. 129). Allen et al. focused on the benefits for protégés and found that there were advantages in career progression as well as social support and job satisfaction. In a similarly conducted meta-analysis that focused on the benefits for mentors, Ghosh and Reio (2013) found that job satisfaction, organisational commitment, job performance, and career success were benefits for the mentors. In both of these meta-analyses, assistance relating to career success, sponsorship, psychosocial support, and role-modelling were all part of successful mentoring relationships. In a meta-analysis of youth, academic, and workplace research on mentoring, Eby et al. (2013) noted that the kind of the learning varies according to the nature of the mentoring relationship, but found commonalities across the areas in protégés' positive perceptions of the instrumental and psychosocial support they received

Varieties of Mentorship

In her analysis of mentoring in business and industry, Gibson (2004) drew attention to "a continuum of roles that range from providing support and advice to highly developmental" (p. 266). Gibson referred to "non-hierarchical (e.g. peer-to-peer) and less exclusive (e.g. group mentoring), as well as additional technology-based alternatives" (p. 266) as extensions of mentoring relationships. Gibson also drew attention to the notion of formal and informal mentoring relationships, according to how the relationship has been organised. The relationship between the mentor and the mentee or protégé can be organised in two ways: either formally (or structured), in which the relationship is mandated, initiated, and sustained by some agency, usually with specified goals, or informally (or unstructured), in which the parties to the relationship initiate and sustain it themselves and goals are not necessarily articulated. Inevitably, the boundary between the two generic forms is fluid, for example when an informal relationship develops into a formal relationship, and vice versa.

Ragins and Cotton (1999) found that informal relationships have the most easily recognised advantages for protégés in terms of career development and psychosocial functions as well greater satisfaction with their mentors. Perhaps this is because they result from the mutual identification of an existing or anticipated shared interest, and thus are more likely to be based on a mutual personal compatibility. Desimone et al. (2014) had similar findings in exploring the mentoring of novice teachers; nevertheless, both groups of authors argued for a mix of formal and informal mentoring. Formal or mandated mentoring ensures that key work-related issues are more certain to be addressed, but research indicated that careful attention must be paid in the selection of partnerships that are compatible.

Although most studies of adult mentor relationships deal with the work-place, aspects relating to success can apply to situations other than work-related environments. This is particularly the case for the informal relationships that characterise most of the kinds of mentoring we found in our case stories. Murray's (2001) reflection is illustrative:

> It is highly likely that the historical examples [she described] of mentoring relationships just happened through some spontaneous interaction of the two people. Either party may have initiated the contact. The protégé may have thought *I'd like to learn more from that person* or may have seen the person do something that had good results; success is seductive. Or the mentor may have been unconsciously attracted to the fresh perspective of the protégé and may have offered a few suggestions for furthering the protégé's work. No matter who initiated the first exchange, both parties undoubtedly felt good about the results of that interaction and sought additional contact. The seeds of a productive relationship were sown in fertile ground, while shared achievement continued to strengthen the bond.
>
> *(pp. 20–21, original emphasis)*

Whether structured or unstructured, formal or informal, there is much evidence in the literature of the value of mentoring. It is not coincidental that many of our case stories feature it, either through identifying a suitable mentor or role model, or by becoming one.

Learning Relationships in Our Case Stories

Nearly all of our case story authors recounted learning from other people in one way or another. Most often this was a face-to-face relationship, but not all would be described as mentorship. Of our case study authors, Hugh, who was working in a rapidly evolving and technology-based industry, likely had the most structured work-place mentoring. His progression through the ranks of aircraft engineering involved numerous training courses, and he acknowledged considerable mentoring by *"some influential professionals"*, particularly at high levels of the aviation surveying business. As Hugh stated, *"nothing beats experience in the field which can only be gained by getting out there under the guidance of a more experienced Surveyor"*. Hugh also became a mentor himself when, for example, he "*ran a small team repairing damaged aircraft for both BA and other carriers, in the United Kingdom and overseas"*.

Some other authors, like Hugh, had a number of mentoring relationships as they progressed in their chosen work. Warren, Tiki, and Kristen were enabled to learn needed information and skills from others more experienced in the interpretation of heritage, knowledge of plants in a botanic garden, and explaining science centre phenomena, respectively. Keith's employment with a wine company allowed him to learn from a range of interactions with grape-growers, wine-makers, a botanist, and his *"boss"*, who taught him about the industry as Keith *"picked up a lot of bits and pieces along the way"*. Ana's discussions with the nurse who ran the antenatal course she attended provided her with valuable information, a learning relationship that was far more useful to Ana than the antenatal course itself. The role model was not always someone new: Warren paid tribute to his Uncle Bert, as a scientist who was a major influence on his activities.

Having mentoring relationships become long-term learning relationships was common, and frequently our authors changed their role from mentee to mentor. Mary's early contacts with mothers of other Down Syndrome children were useful while she was developing knowledge of her own child. She then became a mentor to other parents as a leader in the local Down Syndrome Association. Such associations might be thought of as self-help groups but, like the hobby clubs, they are mutual learning partnerships with considerable fluidity as members move in as novices and later become mentors themselves. During Michael's silver-smithing technical course he had a mentoring relationship with the tutor, although it seems clear that as time went on, the tutor was also learning from Michael as he found ways to create more complex pieces. When the course closed, Michael and some of his fellow students formed a club that was successful, at least in part, because the members were all helping each other. This was peer-to-peer mentoring, although the members themselves might not have considered it a mentoring relationship, just a cooperative help-each-other approach to their jewellery making. However, these informal mentoring relationships are important, and are often long-term learning partnerships, particularly to hobbyists like Michael and also Richard, who works with his club members on a range of restoration projects. In other examples of a peer learning partnerships, Warren's group of interpreters in New Zealand are mutually supportive and they continue to learn from each other. Paulette learned some of her early geological knowledge about the Mere Fault by walking across the downs with geologically-minded friends and discussing what they were seeing. She now participates in a U3A group leading geology field-trips herself.

Liz's mentorship by the instructor of the "Science for Mothers" course extended beyond participation in the course because the instructor was available to be contacted later for additional learning help. Clearly, Liz was also participating in parent–child mentoring with regard to science (and doubtless other) learning. Paul was the other parent whose story involved mentoring his child, and he was the only one not to mention learning from another person; apart from the museum display of dinosaurs, any new learning came from online resources. Indeed, online resources became increasingly important in most of our case study authors' learning.

Learning Relationships and Online Media

As foreshadowed in Gibson's (2004, p. 266) comment that technology-based alternatives extend the range of mentoring relationships, our case stories demonstrate a range of learning relationships supported by the Internet. Of course, the Internet was a ready source of knowledge, as in Paul's search for dinosaur information, Warren's use of it for *"quick updates"* on heritage information, and Tiki's plant identifications. The Internet was even the source of the podcasts about pseudoscience that Ketan listened to while driving. While readily assessable and easy to search, in terms of communicating information, the Internet is like printed media: it is one-way communication of knowledge transfer. In contrast, social media and web-based learning platforms are different because they support two-way communication.

The use of new media has been discussed in Chapter 9 and here we reiterate that it has two great advantages for the self-directed learner: it is available at any time, and the learner can set the pace and depth of learning. When the learner is accessing a learning platform such as an online course, or an interactive site in general, then it is possible to have structured support that can be considered a mentoring relationship, but one where the mentor and mentee do not have a face-to-face relationship. As we write, we are aware that many formal courses are now being offered online instead of face-to-face, but their success is dependent on having trusting relations between learners and more knowledgeable others. As

Hamburg (2013) pointed out, "a constant presence of experienced and qualified mentors in the Web-based platform is required. The platform should support motivation and retain students in the learning process and [be] a real mentoring and not be understood as a supervisory tool" (p. 220). Although Hamburg was referring to mentoring in set courses, designers of Internet-based learning platforms must be able to offer feedback and encouragement to the learner if they are to persist with the experience.

This aspect was evident in Tina's story. Her discovery of the unusual bumblebee she identified by an Internet search has evolved into participation in citizen science projects that are managed almost entirely on the Internet, with additional resources for participants such as newsletters. However, Tina's learning increased when opportunities were provided for feedback via an online identification tool and her ability to participate in conferences allowing face-to-face discussion with experts. Tina stated *"the combination of on-line interaction with some opportunity to meet other people interested in the topic is ideal for me"*. Liz also became involved in Galaxy Explorer, an online citizen science project than took a "crowd sourced" approach to get help in identifying over 200,000 galaxies. Although communication was essentially one-way – Liz contributing her findings to the data base – Liz learned how to classify the galaxies and found it *"interesting looking at the different types of galaxies, and thinking about how far away they are, and how amazing our universe is"*.

The critical role of feedback is clearly demonstrated by one of our own experiences. One of us (LR) has established a mentoring relationship by email with a curator at a large museum. After finding an interesting unknown fish washed up on a beach, she emailed a photograph to an ichthyologist whose contact details were available on his museum website. Because of his continued willingness to assist her with identification and encouragement about the value of her observations, several years and many fish later, she has joined a regional citizen science project using an Internet platform to map fish species and their changing distribution due to climatic warming of the oceans. Although all contact has been by email, this mentor's help and enthusiasm enabled her not only to gain scientific knowledge about fish, but also to contribute to a worthwhile science project that puts her in touch with a range of other fish scientists and interested enthusiasts.

We see that relationships with other people are easy to initiate and can develop into valuable learning partnerships for the self-directed learner. Through social interactive platforms, including blogs, wikis, and online forums, learning relationships may develop that are more like peer interactions than mentoring. Ketan's extensive use of blogging is part of his *"day job [that] involved the digestion and dissemination of scientific information"* from all kinds of media, printed and otherwise. Ketan is now a science communicator in his own right, and certainly a source of information to other learners.

How to Support Self-Directed Learners

Our case story authors accessed a variety of static resources in their learning, such as using the Internet or printed media for information. Some ideas to make these resources more accessible were given in Chapters 8 and 9. In this chapter we are focusing on how learners can be supported by other people to develop the knowledge and skills needed to progress their learning. Who are these people and where might they be found?

Likely Supporters of Self-Directed Learners

A report from the Center for Advancement of Informal Science Education (CAISE, 2016) synthesised a list of the resources for outreach in STEM education and provided examples of their operation. The report began with this statement:

> Over the past ten years, investments in infrastructure for informal STEM education and science communication have resulted in significant growth in the number and variety of resources and depth of expertise available to members of the STEM research community wishing to develop outreach, engagement and broader impacts [*sic*] activities.
>
> *(p. 4)*

CAISE offered a "partial list" of the types of experiences available that included the following:

> Informal STEM education institutions, including science and technology centers, natural history museums, visitor centers connected to National Labs/Large Facilities, botanical gardens, nature centers, park visitor centers, zoos, aquaria, and planetariums; Media, e.g. television, radio, film, social media, science journalism; Youth, community and afterschool programs, e.g. out-of-school and youth development programs; Citizen science organizations, events, and programs; Science festivals, cafes, and other live public events; Cyberlearning platforms; Libraries; Adult-serving organizations; Environmental and conservation-related organizations; Health-related organizations.
>
> *(CAISE, 2016, p. 5)*

This partial list demonstrates the enormous variety of resources for learning science and technology that exist in many countries of the world; a list so big that it is not possible for us to consider each resource separately and recommend how its participants might best offer support to the adult learners who seek them out. Grouping these resources will allow us to offer generic suggestions about how to support adult learners, then illustrate this with more detailed examples.

Reviewing the discussion in Chapter 8, including the contents of Table 8.1 which lists clusters of resources, we see that our case story authors had various kinds of learning relationships with other people. Many of our authors consulted with experts in the field. A number encountered teachers, who were sometimes mentors or role models, or other educators who worked in educational institutions like museums. As well, most of our authors formed learning relationships with peers, friends, and colleagues, particularly those who belonged to clubs and associations, relationships that involved sharing experiences and helping each other. With this variety in relationships in mind, together with the list of resources above, we suggest that there are three broad groups of people who are likely to be involved in supporting adult learners to engage with science and technology.

1. Educators providing formal learning experiences about science and technology for adults through courses and workshops. Many of these educators are technical school teachers, university lecturers, tutors in adult education, or outreach/communication officers in industrial organisations.
2. Specialists and community liaison people working in organisations that provide information to the public about science and technology issues, including aspects relating to public policy, health matters, or the environment. These organisations may be governmental or voluntary.
3. Staff in the educational sections of cultural organisations, for example, libraries, museums, science centres, zoological and botanical gardens, aquaria and observatories, environmental and heritage sites and centres.

If these people are those most likely to have opportunities to support self-directed learners, how might they best go about it? Early in this chapter we suggested that self-directed learners needed four clusters of skills associated with four steps in the learning process: Learners must decide on the information needed, access that information, make the best use of the information, and reflect on what was being learned. We also identified three essential attributes or personal resources to ensure success in achieving their learning goal: Learners must have positive motivation towards the task that has been chosen, active engagement in the process of learning, and a positive belief in their self-efficacy as a learner. In the next section, we consider each of the three groups of people above with reference to these essential skills and attributes.

Educators Providing Formal Learning Experiences

Formal learning experiences for adults usually involve courses and one can imagine a continuum of formality, from a structured, curriculum-based course assessed for certification, to a loosely organised workshop, or series of workshops, responding to participants' needs to learn something specific. Some of our authors had formal tertiary qualifications, like Ketan, others did not, but may have participated in technical courses, like Michael. The learning from these courses was variably valuable depending on the relevance of the content. Ketan found his university courses uninteresting until he found one in neuroscience when he *"became truly invested in his studies. Combined with psychology, the degree offered an answer to something that had been bugging me for a long time: why do people reject science, and believe weird things instead?"* At the other end of the course continuum, Michael found his technical course relevant because the tutor was sufficiently flexible to respond to his needs, Michael wrote *"With my tutor's help, I became more confident and more experimental"*, making it clear he was getting the specific assistance he required.

The extent to which educators running courses can assist their learners' particular needs is dependent, in part, on the necessity to follow a set curriculum. It is also dependent on the quality of the instruction. In a guide book for teachers of adults, Wlodkowski and Ginsberg (2017) referred to the "motivating instructor", one who helps adults "to genuinely want to learn" (p. 47) and who demonstrates five essential elements or skills. These skills are expertise in their subject area, empathy for learners to adapt to their needs and level of experience, enthusiasm about their subject matter, clarity in their instruction, and cultural responsiveness by being respectful and responsive to learners.

The guide by Wlodkowski and Ginsberg (2017) for teaching adults was set in the context of formal taught courses that have a curriculum and usually assessment, and there is an extensive volume of literature about how to teach such courses. Our interest, however, is in self-directed learners, who do not need to be motivated per se, but the educator from whom they seek help must have similar skills to the motivating instructor. They must have empathy, in that they listen to the learner to understand their needs, and respect for the current knowledge and experience of the learner. The instructor must understand the content thoroughly and know where to find it, and also understand the learner's context so they are able to offer knowledge and information that is focused and at the cognitive level at which they can cope. Being encouraging and supportive of learners increases their satisfaction in the help they receive.

For the majority of self-directed learners, curriculum-based courses have much content that the learner does not need. For example, the antenatal course attended by Ana had a syllabus, but it was not helpful to her because she already knew the content. Discussion with the instructing nurse after the course was more helpful because she could respond to Ana's questions and supply the information she did need. Self-directed learners are more suited by educators who are sufficiently flexible to adapt to their needs, somewhat like the need for flexible mentors referred to by Goldner and Mayseless (2008) who were able to determine their youthful mentee's needs and respond to them in appropriate ways.

One skill a supportive educator needs is to be able to determine the nature of the self-directed learner's problem. Roberts (2000) cited Wittgenstein (1958) who posited that there were

> two kinds of problems: problems of ignorance – things that exist that we do not know enough about, and therefore require more information – and problems of confusion – we have the information but we do not know what it amounts to. Wittgenstein's advice, it transpires, is that we should seek to clarify the information we have rather than seek and acquire more.
>
> *(Roberts, 2000, p. 145)*

Roberts (2000) used the distinction between ignorance and confusion as a basis for starting his literature review about the mentoring relationship, but we can use it to advise supporters of self-directed learners. Listen to the learner to find out what is the basis for the information needed. For self-directed learners attending courses, quite often the problem relates to the finding of relevant information, so the supportive educator may be able to provide it, or help the learner to find it. If the problem requires understanding of current information, then the solution lies with clarification, a process which invariably involves two-way communication between the supporter and the learner.

Specialists and Community Liaison People

Self-directed learners seeking information from organisations about policy, health, or the environment can turn either to specialists who are expert in their particular field, or to liaison people who tend to provide information (often a one-way process via various kinds of information sheets) or work with the community. In Chapter 8 we discussed how our learners sought help from these people. The most challenging experiences were those involving health professionals and our learners recognised the need to ask the right questions, but sometimes they needed help to focus their questions so they could get the right answers.

The need for effective communication is well recognised in the health professions. There is considerable literature about how doctors can be most helpful to their patients, and the emphasis is always on the importance of effective communication for quality health care. For example, the American College of Obstetricians and Gynecologists (ACOG, 2014) has an endorsed Committee Opinion on the approaches their doctors should take to ensure effective communication with patients. ACOG stated that "developing effective patient–physician communication requires skill in conducting patient-centered interviews; conversing in a caring, communicative fashion; and engaging in shared decision making with patients" (p. 2). Travaline, Ruchinskas, and D'Alonzo (2005) argued that "the manner in which a physician communicates information to a patient is as important as the information being communicated" (p. 13). Travaline et al. recognised that communication skills take time to develop and ongoing practice, but result in better clinical outcomes. They suggested nine practical steps to help physicians strengthen their personal communication style: (1) Assess what the patient already knows, (2) Assess what the patient wants to know, (3) Be empathetic,

(4) Slow down, (5) Keep it simple, (6) Tell the truth, (7) Be hopeful, (8) Watch the patient's body and face, (9) Be prepared for action (pp. 15–16). These authors also listed some communication "traps to avoid", including the use of highly technical language and jargon, not listening to the patient, not showing appropriate concern, or not verifying that the patient has understood the information. We recognise these as impediments to effective communication more generally.

A commentary in the *Wall Street Journal* (April 12, 2013) provides interesting reading and an apt summary of how experts can help learners. In response to an article relating to poor doctor–patient communication, the *Journal* sought the opinions of a number of medical experts about the most important skill needed to improve doctor–patient communication. The answers were pithy and sensible, but central to them all was the need to pay attention to the patient's needs. This advice is entirely applicable to any expert who seeks to inform and support learners in the field. The persistence of this theme in the health education literature and elsewhere shows that communication skills – however described – are valued in the profession but that they are difficult to acquire. Effective two-way communication is fundamental to any learning relationship.

Staff in the Educational Sections of Cultural Organisations

It is difficult to find a single term for the staff of cultural institutions whose role it is to assist visitors or users of their facilities. There are librarians, curators, docents, explainers, ground or floor staff, guides, leaders, and many other titles. There are sufficent commonalities in their roles to make do with one term because all are charged with assisting and informing those who seek their help. We will refer to them as supporters because they are there to assist visitors to gain from their experience.

Useful background information for supporters can be found in *The Museum Experience Revisited* (Falk & Dierking, 2013), a handbook written for staff at these institutions based on several decades of research into the factors that affect the visitors' experience. Falk (2009) described seven kinds of identity-related motivations for adults visiting museums – explorers, facilitators (of other visitors' experience, such as children), professional/hobbyists, experience seekers, rechargers, respectful pilgrims, and affinity seekers – titles which conjure up descriptions of the motivational reasons people may have for their visit. Supporters will be aware of this range of motivations. Many visitors to such institutions may not be on a mission to learn, they may be browsing exhibits or bookshelves, or helping their children to enjoy a learning (and/or entertaining) experience, or simply enjoying the ambiance of the institution. This means that sometimes supporters can best help by recognising that visitors may not want to interact, and leave them alone to pursue their own reason for visiting.

Caro (2003) described how museums of various kinds are ideal places for learning, particularly by autodidacts, who can wander at will and are free to help themselves to whatever knowledge they might seek, be it particularly sought for or obtained incidentally. Supporters need to be sensitive towards visitors who do not want to talk. When visitors do ask questions, Morris (2015) pointed out that supporters can develop this opportunity for interaction by attentive listening and sharing, rather than telling, information. Morris also warned that curiosity is quickly dampened by giving immediate direct answers, or answers with too much information or jargon.

Whether or not a learning relationship is created between a self-directed learner and a supporter in education institutions depends on the learner's reason to visit. Paul, who took Emma to the Natural History Museum to see the dinosaurs, does not mention any interaction with staff so viewing the exhibits and their signage may have been sufficient for their purpose. When interaction is initiated either by the learner or supporter, the first step for the supporter is to determine what kind of support should be offered. To use Wittgenstein's dichotomy (as cited by Roberts, 2000), do learners have a problem of ignorance about the issue, or a problem of confusion that they need help to sort out? The answer will determine how best supporters can assist.

Often self-directed learners first need help to find the information they want and they may then need help to make best use of it. Supporters assist by being intermediaries between the learner and the information. Librarians are key resource people in this area, so let us use them as an example. To be helpful they must do two things. First, they must listen to the person to ascertain what is required, and second, they must know not only their own collections, but know of, and understand, other sources and how to access them. Listening is important because it enables the librarian to assess what the learner knows already, what cognitive level is appropriate, and what format the learner prefers to access. Careful questioning by the librarian will help to focus the search and lead more quickly to the appropriate resource at the appropriate level. Some librarians refer to this as a "reference interview" (M. Coen, personal communication, March 28, 2017) that is fundamental to being of most assistance to the learner. Second, finding appropriate resources requires deep familiarity with the scope and content of the library shelves, or other places (including online) that can assist the learner. Librarians know that for cutting edge knowledge, even new books may be out-of-date due to publication delays, and that the most current source of information may be in some other format that they can assist the learner to access. There is considerable difference between the extent and quality of resources that can be accessed in an institutional library compared to a public library, and the local librarian will be able to redirect learners as required. Sometimes the librarian's assistance may simply be to help learners find their way through the online catalogue. Use of these can be frustratingly different in different libraries, so clear instructions alongside the search computers are very helpful, but empathetic librarians are even more so!

For those staff working in environments more devoted to interpretation than librarians, supporting learners requires different kinds of assistance because they may be focusing on a specific exhibit or feature of the learning environment. A key role for staff is interpretation of that exhibit, feature, or environment, so staff need a good understanding of those things, but also need to understand what the learners are seeking. Considerable research has indicated that staff in such roles are often didactic in their interactions with learners (King & Tran, 2017), rather than taking a more responsive approach to learners' actual needs that may assist their learning. Considerable attention is now being given to assisting staff to develop reflective practice (see Patrick, 2017, for examples of such programs). Ash, Lombana, and Alcala (2012) reported the results of research encouraging museum educators to change their practice to become more reflective and to listen more to visitors. The following rather long quote about Norman, one of the museum educators, is instructive.

> Initially educators like Norman thought they needed to control and assert their power as museum authorities, by teaching, telling, lecturing, and controlling interactions. The irony was that by using such tactics, they often lost control of the interaction and blocked the learning process, as visitors felt intruded upon, intimidated and/or ignored in what they had to say or how they said it. The paradoxical surprise was discovering that by letting go of controlled interactions, educators were more able, rather than less, to affect a positive outcome. This phenomenon truly surprised the educators in this cohort.
>
> *(Ash et al., 2012, p. 38)*

Perhaps if a mantra were needed to enable supporters to assist learners it would be the signs at railroad crossings: stop, look, and listen. Stop talking, look at the learner, and listen to what they want to say. Then you can choose the most effective way to be supportive.

Effective Communication

The commonality in all of the effective learning relationships we have described in this chapter is successful two-way communication. Effective supporters, no matter what their role, are effective communicators, and this requires mastery of a series of broad skills. Stocklmayer and Rennie (2017, pp. 541–542) set out the skills for effective two-way communication and those we see as critical to the supporters of self-directed learners are as follows:

- The ability to match the communication to the beliefs, values, and knowledge of the learner.
- The ability to "translate" and present the science and technology in a clear, jargon-free narrative, incorporating all the principles of good communication.

- An understanding of how social contexts affect the learning of science, including the myriad cultural and psychological influences that affect people's perception of the science in their individual environments. The ability to modify and change the communication to match these changing contexts.

Readers will recognise consistency with the essential skills of Wlodkowski and Ginsberg's (2017) "motivating instructor", but here the emphasis is on two-way communication. Unlike the student–teacher relationship in taught courses, adult self-directed learners are more likely to be equal partners in the learning relationships we have focused on here. As Morris (2015) pointed out, the "tutor needs to swallow hard and welcome the topics as laid down by the participants ... the key is to become a careful listener, to try to understand what lies behind the presenting question" (p. 85). The desired outcome is to enable the self-directed learner to take appropriate action and to learn the relevant concepts or skills required.

It is imperative that learners are encouraged to retain a positive attitude to the task of learning and persist in solving the problem or issue that needs resolution. It is important to remember Eraut's (2000) statement about the importance of challenge. We noted this in Chapter 6 and repeat it here: "if there is neither a challenge nor sufficient support to encourage a person to seek out or respond to a challenge, then confidence declines and with it the motivation to learn" (Eraut, 2004, p. 269). An additional and significant outcome is that a successful learning experience not only contributes to the learner's positive self-efficacy and willingness to pursue further issues, but also contributes to the development of knowledge about how they prefer to learn, making their next learning adventure into science and technology more efficient. Finally, in addition to solving the immediate challenge being addressed, learners also progress towards becoming scientifically literate. This is the theme we take up in Chapter 11.

References

Allen, T. D., Eby, L. T., Poteet, M. L., Lentz, E., & Lima, L. (2004). Career benefits associated with mentoring for protégés: A meta-analysis. *Journal of Applied Psychology*, 89 (1), 127–136.

American College of Obstetricians and Gynecologists. (ACOG). (2014). Effective patient–physician communication. Committee Opinion, Number 587. *Obstetrics and Gynecology*, 123, 389–393.

Ash, D. B., Lombana, J., & Alcala, L. (2012). Changing practices, changing identities as museum educators. In E. Davidsson & A. Jakobsson (Eds.), *Understanding interactions at science centers and museums – Approaching sociocultural perspectives* (pp. 23–44). Rotterdam, The Netherlands: Sense Publishers.

Bandura, A. (1977). Self-efficacy: Toward a unifying theory of behavioral change. *Psychological Review*, 84(2), 191–215.

Caro, P. (2003). The autodidactic museum in France and other countries. In J. Solomon (Ed.), *The passion to learn: An inquiry into autodidactism* (pp. 106–114). London: RoutledgeFalmer.

Center for Advancement of Informal Science Education (CAISE). (2016). *Informal STEM education: Resources for outreach, engagement and broader impacts.* Washington, DC. Retrieved from http://www.informalscience.org/sites/default/files/CAISE_Broader_Impacts_Report_2016_0.pdf

Deci, E., & Ryan, R. (1985). *Intrinsic motivation and self-determination in human behaviour.* New York: Plenum Press.

Deci, E., Vallerand, R., Pelletier, L., & Ryan, R. (1991). Motivation and education: The self-determination perspective . *Educational Psychologist*, 26(3), 325–346.

Desimone, L. M., Hochberg, E. D., Porter, A. C., Polikoff, M. S., Schwartz, R., & Johnson, L. J. (2014). Formal and informal mentoring: Complementary, compensatory, or consistent? *Journal of Teacher Education*, 65(2), 88–110.

Eby, L. T. d. T., Allen, T. D., Hoffman, B. J., Baranik, L. E., Sauer, J. B., Baldwin, S., ... Evans, S. C. (2013). An interdisciplinary meta-analysis of the potential antecedents, correlates, and consequences of protégé perceptions of mentoring. *Psychological Bulletin*, 139(2), 441–476.

Eller, L. S., Lev, E. L., & Feurer, A. (2014). Key components of an effective mentoring relationship: A qualitative study. *Nurse Education Today*, 34(5), 815–820.

Eraut, M. (2004). Informal learning in the workplace. *Studies in Continuing Education*, 26, 247–273.

Falk, J. H. (2009). *Identity and the museum visitor experience.* Walnut Creek, CA: Left Coast Press.

Falk, J. H., & Dierking, L. D. (2013). *The museum experience revisited.* Walnut Creek, CA: Left Coast Press.

Fredricks, J. A., Blumenfeld, P. C., & Paris, A. H. (2004). School engagement: Potential of the concept, state of the evidence. *Review of Educational Research*, 74(1), 59–109.

Ghosh, R., & Reio, T. G.Jr. (2013). Career benefits associated with mentoring for mentors: A meta-analysis. *Journal of Vocational Behavior*, 83(1), 106–116.

Gibson, S. K. (2004). Mentoring in business and industry: The need for a phenomenological perspective. *Mentoring and Tutoring: Partnership in Learning*, 12(2), 259–275.

Goldner, L., & Mayseless, O. (2008). Juggling the roles of parents, therapists, friends and teachers – a working model for an integrative conception of mentoring. *Mentoring & Tutoring: Partnership in Learning*, 16(4), 412–428.

Hamburg, I. (2013). Facilitating learning and knowledge transfer through mentoring. *Proceedings of CSEDU 2013–5th International Conference on Computer Supported Education* (pp. 219–222). Retrieved from https://pdfs.semanticscholar.org/8aeb/3221ce4680ea2e4ce8c8e78fad6f9d721f13.pdfhttp://dx.doi.org/10.1037/a0029279

Illeris, K. (2002). *The three dimensions of learning.* Leicester, UK: Roskilde Universitet.

Jarvis, P. (1987). Meaningful and meaningless experience: Towards an analysis of learning from life. *Adult Education Quarterly*, 37(3), 164–172.

King, H., & Tran, L. (2017). Facilitating deep conceptual learning: The role of reflection and learning communities. In P. G. Patrick (Ed.), *Preparing informal science educators: Perspectives from science communication and education* (pp. 67–85). Cham, Switzerland: Springer Nature.

McKinnon, M., & Vos, J. (2015). Salvaging science literacy: Engagement as a "threshold concept" for science education and science communication. *International Journal of Science Education, Part B,* 5, 297–318.

Merriam, S. B., Caffarella, R. S., & Baumgartner, L. M. (2012). *Coursesmart: Learning in adulthood: A comprehensive guide*. Hoboken, NJ: Jossey-Bass.

Metzger, M. J., & Flanagin, A. J. (2013). Credibility and trust of information in online environments: The use of cognitive heuristics. *Journal of Pragmatics* 59, Part B, 210–220.

Morris, A. (2015). *Getting to grips with science: A fresh approach for the curious*. London: Imperial College Press.

Murray, M. (2001). *Beyond the myths and magic of mentoring: How to facilitate an effective mentoring process*. San Francisco, CA: Jossey-Bass.

Patrick, P. G. (Ed.) (2017). *Preparing informal science educators: Perspectives from science communication and education*. Dordrecht, The Netherlands: SpringerNature.

Ragins, B. R., & Cotton, J. L. (1999). Mentor functions and outcomes: A comparison of men and women in formal and informal mentoring relationships. *Journal of Applied Psychology*, 84(4), 529–550.

Roberts, A. (2000). Mentoring revisited: A phenomenological reading of the literature. *Mentoring & Tutoring*, 8(2), 145–170.

Ryan, R., & Deci, E. (2000). Intrinsic and extrinsic motivations: Classic definitions and new directions. *Contemporary Educational Psychology*, 25, 54–67.

Solomon, J. (Ed.). (2003). *The passion to learn: An inquiry into autodidactism*. London: RoutledgeFalmer.

Stocklmayer, S. M., & Rennie, L. J. (2017). The attributes of informal science education: A science communication perspective. In P. G. Patrick (Ed.), *Preparing informal science educators: Perspectives from science communication and education* (pp. 527–544). Cham, Switzerland: Springer Nature.

Travaline, J. M., Ruchinskas, R., & D'Alonzo, G. E.Jr. (2005). Patient–physician communication: Why and how. *Journal of the American Osteopathic Association*, 105(1), 13–18. Retrieved from http://jaoa.org/

Wall Street Journal. (2012, April 12). The experts: How to improve doctor–patient communication. Retrieved from https://www.wsj.com/articles/SB10001424127887324050304578411251805908228

Wlodkowski, R. J., & Ginsberg, M. B. (2017). *Enhancing adult motivation to learn: A comprehensive guide for teaching all adults* (4th ed.). San Francisco, CA: Jossey-Bass.

Yun, J. H., Baldi, B., & Sorcinelli, M. D. (2016). Mutual mentoring for early-career and underrepresented faculty: Model, research, and practice. *Innovative Higher Education*, 41 (5), 441–451.

Zimmerman, B. J. (2000). Self-efficacy: An essential motive to learn. *Contemporary Educational Psychology*, 25, 82–91.

11

ADVANCING THE CAUSE OF ADULT LITERACY IN SCIENCE AND TECHNOLOGY

As we illustrated in the introduction to this book, science and technology permeate the ways we live, work, and spend our leisure time. Leading fulfilling lives depends on having some awareness of science and technology, and feeling sufficiently comfortable with them to be able to identify relevant information and use it to make decisions about the issues and problems that arise in daily life. In Chapter 1 we rehearsed the calls for a science and technology education that contributed to the preparation of all citizens for every aspect of their lives. Even in cases where courses are initially relevant to everyday life, the march of progress in science and technology will make such preparation less relevant as the years of adulthood pass. But the need to know something about science and technology will not pass. Indeed, as adults living independently in the world, and often responsible for the welfare of others, it will become ever-more important to know and understand the science and technology underpinning daily objects, events, and issues. But, as we (Stocklmayer, Rennie, & Gilbert, 2010) and others (for example, Aikenhead, 2006; Fensham, 1985, 2008) have pointed out, most often, the school science curriculum does not succeed in providing its graduates with sufficient understanding of science to achieve this outcome. Why might this be?

Science in the School Curriculum

The school curriculum began to include the subject of "science", in some form, in Western Europe and the United States towards the end of the nineteenth century and in other areas somewhat later (Jenkins, 1994). The initial intent was to identify those students with an aptitude for science or engineering who would later benefit from higher education in these fields. The curriculum was broad, within the confines of research knowledge of science at the time. It consisted of

DOI: 10.4324/9780429434303-11

the development of an understanding of core, usually abstract, concepts, together with practical experience of phenomena that were considered important. As time went by, not only did the volume of knowledge in these fields expand exponentially, placing increasingly greater demands on the curriculum for these specialist students, but the range of motivation of students altered. An ever-growing proportion of students, those for whom science was increasingly a mandated rather than a voluntary subject, had no intention of continuing to study science or engineering. They might, however, be interested in the applications and implications of science as these had resonance with the rest of the curriculum.

The way that the science curriculum for these two different but overlapping groups of students was organised varied from country to country. In some countries, such as the United States, discrete courses in physics, chemistry, and biology were provided in separate school years, with students only being required to study a minimum one of them in order to matriculate. In others, such as the United Kingdom, the science curriculum was broken down into "modules", each within one of these three subjects, which were orchestrated into courses that all students were required to take. In both of these types of curriculum arrangement, the presumed needs of the "pre-specialist" students predominated. In many countries, subject specialist courses were retained and "general science" courses were provided for the non-specialist majority. However, from the 1970s onwards, the growing recognition of the importance of science and technology in personal, social, and economic life led to the adoption of "national curricula". A module-structured form of curriculum became fairly universal, but now with graduated "levels of attainment" included in order to accommodate the range of student interest and commitment. In some countries these modules are now strictly mandated for all students as in the United Kingdom, but elsewhere, for example, in the United States, they remain "advisory" at the moment.

The essential nature of the most recent of these trends in the school science curriculum has been encapsulated in the seminal paper by Roberts (2007) referred to in Chapter 1. He described science curricula overall in terms of two "Visions", where, to recap, Vision I is characterised by "looking inwards at the canon of orthodox natural science, that is, at the products and processes of science itself" (p. 730). Vision II is obtained by looking outwards, considering "the character of situations with a scientific component, situations that students are likely to encounter as citizens" (p. 730). Much of the research literature about the learning of science has dealt with the importance of engaging school-age students in formal educational science contexts, either Vision I or Vision II, in order for meaningful learning to occur. In the case stories described in this book, however, engagement was a self-imposed necessity for achieving the desired goals, which seemed to cut across the two Visions, and it came about because of strong motivation to pursue those goals. For our case stories, Vision II was the key to the desired achievement. From this we believe that it reflects the importance of a Vision II approach in the education of all citizens.

Roberts and Bybee (2014) later revisited these two Visions in the context of changes in curricular approaches world-wide. They reflected on the seeming inability of science curricula to address Vision II, and presented evidence that there was a trend of reversion to Vision I. It seems to us self-evident that it is within Vision II that the scientific literacy of all citizens has the possibility of development. Further, traditional Vision I science curricula, however coated in a cloak of "STEM", are not able to provide this development, for their purpose is largely, if not completely, concerned with the exploitation of science in the economy. Moreover, Roberts and Bybee (2014) firmly challenge the notion that simply "knowing science", the Vision I approach, enables a person to apply that science in life situations.

> As a matter of policy, however, we want to be quite clear that this represents a choice of Vision I over Vision II – not simply the hope that Vision II goals can be accomplished through Vision I practice. Instead, this choice eliminates a vital part of Vision II, namely an understanding of practical reasoning about science-related personal and societal issues.
>
> *(p. 550)*

From the perspective of scientific literacy, there is a strong trend for recent curricula to retreat from Vision II. To some extent, this retreat may be the consequence of the fate of all curriculum innovations. It may therefore be explained in terms of the notion of "curriculum representation" as suggested by Goodlad (1994) and van den Akker (1998). Any curriculum innovation is, on the Goodlad/van den Akker scheme, subject to six successive but overlapping representations with the passage of time.

1. The *ideal* curriculum phase exists as the original ambition of the originators. In our case this would be the philosophy underlying a Vision II curriculum.
2. The *formal* curriculum phase is represented by the prescribed curriculum document.
3. The *perceived* curriculum phase takes the form as understood by teachers. Their understanding will be framed by the teaching they experienced, which will most likely be of Vision I curricula.
4. The *operational* curriculum phase takes the form that the teacher adopts in the classroom. This will be heavily influenced by what the teacher has previously found the students to expect, namely Vision I.
5. The *experienced* curriculum phase takes the form of what the students understand.
6. The *attained* curriculum phase consists of what the students have actually learned, as reflected in the nature of and their responses to assessment tasks.

The consequences of this retreat during the passage from "ideality" to "attainment" for Vision II curricula, is signalled in the internationally very influential

2012 *K-12 Framework for Education* in the United States that suggests that aspects of Vision II belong in the curriculum for Social Studies. Roberts and Bybee (2014) stated: "This document is an example of withdrawing from Vision II by purifying science education policy through purging the attention to personal and societal perspectives" (p. 550). Further, the Programme for International Student Assessment (PISA), which is such a strong driver of science curricula and educational practice, modified its conception of scientific literacy in 2015. The PISA definition now encompasses explaining phenomena scientifically; understanding scientific enquiry; and interpreting scientific evidence. Aspects such as making decisions about the natural world are no longer present. Regarding PISA, Roberts and Bybee (2014) "propose that scientific literacy, as expressed in Vision II, has reduced emphasis in PISA 2015 compared to both the versions that were assessed in 2000 and 2006" (p. 552). They stressed that, in real life, "Citizens encounter science not in its pure form but in personal, local and global contexts and life situations" (p. 553). Indeed, this is what our case stories have conveyed throughout this book.

It is the view of Roberts and Bybee (2014), however, that Vision II may present classroom challenges that result in Vision I being the default position when it comes to curriculum implementation. Teachers may find teaching the everyday science perspectives required in Vision II either too complex to be taught in the time available or too broad in scope. Vision II may require teachers to move into a more uncomfortable frame than offered by traditional curricula (p. 555). Critically, the authors conclude that "there is more need than ever to develop ways to balance science literacy (Vision I) and scientific literacy (Vision II) in science education programs that can successfully meet the needs of all students" (p. 556).

Meeting the needs of all students will require that links be made between science in school and science in community and there is considerable evidence that teaching science in this way does indeed present challenges for teachers, because it requires a new way of thinking about the nature of science teaching. Moving from a Vision I perspective to a Vision II inclusion of community links often means the development of additional knowledge and skills for the teachers (Rennie, 2011). The mandatory requirements of the now-ubiquitous "national curricula" leave teachers with little, if any, time or opportunities for professional innovation. Nevertheless, developing attitudes to science aligned with Vision II and the knowledge and skills associated with their exemplification in everyday life must be integrated within the presentation of science in schools, despite the limitations of curricular time.

Dealing with Science and Technology in Everyday Life

In this book we explored how adults dealt with matters relating to science and technology in their everyday lives. The narrative case stories presented in Chapters 3 to 6 provide illustration that adults use considerable initiative in exploring a chosen scientific theme. Our case story authors wrote about their lives as lived in the real world, where the issues that concerned them were frequently interdisciplinary; nothing was identified as just "science". Of our authors, only Ketan wrote about science per se, but he linked science and technology together in a way that recognised the value laden context in which they are used.

> *Crichton's warnings of scientific overreach (framed in bouts of gory drama) only furthered my enthusiasm for science and technology. It established, in my mind, that technological expansion through science ought to be a cautious equilibrium, not a blind push for progress.*

Ketan's story is that of a man highly literate in science and technology, whose life is spent thinking about science and pseudoscience, and working in renewable energy and climate change, two of the most pervasive socio-scientific and controversial issues of today. Major world problems, such as dealing with an outbreak of a potentially pandemic virus, such as Ebola or Zika, are not tackled by scientists alone, whole teams of various disciplinary experts are involved to work on a common problem, and at every turn, there are economic, political, social, and cultural values to be negotiated. While everyday life for most of us involves science-related problems much less major than pandemic virus outbreaks, we are still dealing in a world where science is integrated seamlessly with many other subjects and impinged by the values of our personal world.

There is a stark contrast between how the adults in our case stories used science and how science is taught in school. As our case story authors revealed, and Rennie, Venville, and Wallace (2012a) pointed out, in everyday life there is rarely a distinction made between science and technology. It is only the school curriculum where traditionally, and artificially, science and technology are timetabled as separate subjects.

Technology in the School Curriculum

Technology as a discrete subject in the school curriculum (as distinct from applied subjects such as woodwork and domestic science) has a more recent history than science, finding a relatively permanent place in most countries late in the last century (see Donnelly, 1992, for a critical historical review). Initially there was a struggle to separate technology from science, because there was a widely held view that technology was really applied science. Gardner (1994) analysed how science and technology were represented in the school curriculum, concluding that the essential difference between the two was one of purpose, as demonstrated in Chapter 1. The discussion of the Vision II approach to science above makes clear that this Vision requires closely linking science with everyday life and this inevitably includes aspects of technology, together with other, social, ethical, economic, and political values. For the adults who are the focus of this book, curricular issues dealing with the separation of science and technology were never relevant.

The distinction made between science and the humanities subjects at school is also artificial, as S8, one of the student science communicators, discovered in reflecting on learning from the Moon Diary assignment. *"I also found that I have equal passion for science and the arts, and that these two fields do not have to exist in isolation but can rather work together to be powerful science communication tools."* S8's remark is reminiscent of Snow's Two Cultures, an artificial division that exists in people's minds rather than the world we inhabit, where the beauty of a rainbow, or a flower, is inseparable from science. Perhaps it is not surprising that some educators wish to include an "A" for Art in to STEM to create STEAM.

The Curricular Relevance of STEM and STEAM

Despite the transdisciplinary nature of our world, integrating science and technology in educational curricula is neither common nor easy, particularly in high schools characterised by unidisciplinary timetabling and subject-specialist teachers. Even when the pedagogy is strong, the curriculum content is presented in silos. In their book *Teaching and learning STEM: A practical guide*, which discusses aspects of innovative teaching at the tertiary level, authors Felder and Brent (2016) unambiguously refer to "STEM disciplines" throughout the book. The disciplines are accepted as separate (as perhaps might be expected at this level) and the book focuses on how to promote learner-centred teaching and active learning techniques within these separate disciplines. STEM skills thus become, to all intents and purposes, synonymous with inquiry learning skills, as described in many school syllabuses. It seems to us that this skill-based approach to STEM in schools is relatively common, but that true integration of the four disciplines is much more difficult.

If attempts to integrate the conceptual content of the science and the technology curricula have been only marginally successful, widespread integration of the STEAM curricula is even more problematic in practice. It will likely prove impossible, particularly at the high school level where the implications for teacher education seem overwhelming. In the case of adult learners outside of school, however, such debates become irrelevant. What needs to be learned will be approached entirely in a cross- and multi-disciplinary context, framed wholly by the learner's needs and skills. For them, the disciplinary divisions are artificial and of no account; consequently, in this book, we will not dwell further on the advantages and problems of learning science in a STEM or STEAM framework. Nevertheless, it is likely that if school curricula advocated a greater emphasis on active learning and inquiry-based learning it would have positive implications for adults' ability to learn in a self-directed way.

The Relevance of an Integrated Curriculum

Although an integrated curriculum has been recognised as an approach more related to students' lives outside of school, in practice it is difficult to achieve. From their years of studying how teachers in schools attempted to introduce integrated curricula, Rennie, Venville, and Wallace (2012a) advocated a "Worldly Perspective" to curriculum that would provide a "balance between disciplinary and integrated knowledge and connection between local and global knowledge" (p. 119). The authors stated their belief that "schools should seek to provide students with the knowledge that prepares them to be responsible adults and sensible citizens" (p. 120). Sadly, at present, this is not usually an outcome of school science curricula.

Of course there are pockets of exemplary efforts to implement STEM and STEAM in an integrated way, usually in elementary schools where teachers and their timetables have more flexibility. Paige et al. (2018), for example, described three fifth grade case studies of a transdisciplinary curriculum based on a futures perspective on the theme of water resources. While the outcomes were positive for students' cognitive and emotional growth, the project involved long term professional development for teachers, assistance from researchers and support for teachers, both financial and material. Rennie, Venville, and Wallace (2012b) included ten case studies of integration across various subject areas, some successful others less so, demonstrating the difficulty for teachers of moving beyond the status quo. The most successful case study involved a small, remote school for Indigenous children, where the entire curriculum revolved around literacy, with English being a third language for these children (Gribble & Rennie, 2012).

These examples suggest to us that the idea of implementing an integrated curriculum that is an ideal blend of the Vision I focus on disciplinary knowledge and the Vision II focus on science in everyday life would convey a Worldly Perspective to students. However, it cannot be mandated by enforced national curricula. As we have seen, many recent curricula are retreating from aspects of Vision II. On the other hand, small and stable learning environments in schools, visionary and supportive leadership from teachers and administrators, flexible timetables, and making links with the community are just some of the factors that may facilitate a Worldly Perspective (Rennie et al., 2012a, p. 117). Roberts and Bybee (2014) identified three desirable aspects of patterns of discourse in the classroom, facilitating both Visions, *viz*: A theoretical reasoning pattern, to establish warranted knowledge; a technological (design) pattern to show its applications; and a practical pattern "appropriate when one intends to arrive at value-laden decisions that impact other people. This discourse included personal and societal perspectives that can be, for example, economic, political, ethical, and moral in character" (p. 548). Rennie et al. (2012a, p. 118) also invoked the three curriculum interests of Habermas (1971) that facilitated the development of a Worldly Perspective. First, "the technical interest … refers to the need to come to a better understanding of the disciplinary techniques, ideas and concepts behind a topic". Second, "the practical interest refers to the manner in which students make personal sense of the integrated topic being studied, how they solve problems and communicate their ideas." Finally, "the critical interest involves students questioning current practices, considering how those practices may be changed" and taking action.

Although different in definition, the parallels are clear. Science in the classroom must be grounded in context, and the context will be different from one class and one school to another. In all of this, there is one finding that stands clearly above all others: the teacher must be able to implement these goals. No matter how good the curriculum (or how bad), if the teacher is not able to enrich the learning with deeply contextualised material, it will fail. We propose, therefore, that instead of unrealistic hopes that curriculum enforcers globally will enable a broad and enlightened approach to science learning, we focus on enabling teachers to address these goals.

This requires a new vision, indeed, for curricula and for teacher professional development. It would be a vision in which teachers are both professionally equipped and professionally courageous to take *any* curriculum and expand it to be balanced and relevant. We have known for many decades that a good teacher can inspire and empower students to take control of their own learning and make it meaningful. Ideally, all teachers would be helped by professional development to make learning science accessible, personal, and a lifelong interest. Further, if curriculum is based on integration with a focus on context, students will be more engaged (because of the greater relevance); more able to transfer their knowledge; more focused on big, important topics rather than unconnected ideas, and more able to direct their own learning. If this approach were taken, it would also be more likely to capture the essence of a liberal education, one that:

> empowers individuals and prepares them to deal with complexity, diversity and change. It [would provide] students with broad knowledge of the wider world (e.g. science, culture, and society) as well as in-depth study in a specific area of interest. A liberal education helps students develop a sense of social responsibility, as well as strong and transferable intellectual and practical skills such as communication, analytical and problem-solving skills, and a demonstrated ability to apply knowledge and skills in real-world settings.
>
> *(Association of American Colleges & Universities, n.d., p. 1/4)*

All sustained curriculum change is gradual and occurs over a long time-frame. It may be realistic to identify some small-scale, achievable changes that would strengthen those aspects of Vision II that would then be included in Vision I curricula. Feinstein, Allen, and Jenkins (2013) have three suggestions. First, increase the number of contexts studied that involve science being used to solve real-world problems. Such contexts would, in many cases, be labelled as "technology" and partially address the STEM agenda. Second, cultivate students' long-term engagement with science by involving them in projects with out-of-school agencies, citizen science projects being one such avenue. Doing so will be challenging for schools, as it will involve forming professional relationships with outside agencies, such as government research agencies, industrial companies, and museums. Third, develop students' capability to judge the relevance and validity of the claims made by science, for example, within Problem-Based Learning and Science-Technology-Society courses. We may then be closer to achieving that Worldly Perspective that enables scientific literacy.

Developing Literacy in Science and Technology

In this book we are championing self-directed learning in the context of developing scientific and technological literacy and, as explicated in Chapter 1, the two are closely connected. The narrative case stories presented in Chapters 3 to 6 provide illustration that individuals use considerable initiative in exploring a chosen scientific theme. As noted, in this endeavour, our case story authors did not separate the dimensions of science, technology, and engineering in their learning. They thought of them all as "science", merged with other disciplines as the need arose. The outcomes of our case stories were that the authors increased their knowledge of matters scientific and technological. How much, however, did this knowledge-seeking activity extend to a more general approach to "science", as they interpreted it? How much did our authors expand their exploration so as to embrace broader elements of "scientific literacy"? In some cases, their new knowledge was restricted to the narrow focus of their immediate concerns, but, in others, a wider application and interest stimulated a further exploration.

In Chapter 1 we stated that scientifically literate people demonstrate five characteristics. They are interested in and understand the world around them; they engage in discourses of and about science; they are able to identify questions, investigate, and draw evidence-based conclusions; they are sceptical and questioning of claims made by others; and they are able to make informed decisions about the environment and their own health and well-being.

These characteristics all apply, to a greater or lesser degree, to all the case study authors. All identified the questions they needed to ask, even though these may not have been evident at the outset of their enquiries. Their search for knowledge about science led to understanding and, in many cases, further interest in pursuing it in new directions. This is often termed "lifelong learning"; a desirable process for increasing *scientific literacy* over a person's lifetime. Further, they engaged in discourses with various friends and others about science. Often, this led to informed decisions about the environment or health and well-being.

Ana, for example, extended her immediate search for information about her own health to an exploration of what factors might affect the health of her baby.

She continued to learn about the baby's heath after her birth, and will undoubtedly keep to this pattern in the future. We suggest that this progression is a common story in new mothers' experiences. Ana's scientific background led her to be systematic and methodical in her search for information but this process of ongoing learning is common to many mothers of a firstborn child. We suggest that there is scope here for assistance to new mothers from many fields as well as obstetrics and gynaecology; for example, nutrition. Exercise and other aspects of health also have a place in this story. Yet the specific information about what to do and how to do it – and what to understand – remains hard to find and is often incomprehensible.

This was the barrier that Penny faced and it proved a great impediment to wider exploration. The problem of medical jargon and the difficulty of finding information on the web mean that it is very hard to expand one's general knowledge and make informed decisions about one's health. Searching can remain confined to the initial topic and be narrow in focus, which is a less than desirable outcome. This was also initially the case for Mary but she subsequently widened her searches around Down Syndrome to encompass issues of education and health. In all these cases, more accessible information would have been very helpful. Nevertheless, all three of our medical case stories have exemplified the characteristics of increasing scientific literacy as they pursued their learning goal.

For the hobbyists, there is no doubt that ongoing exploration of their special-interest fields widened their knowledge of science and technology and increased their ambitions for what could be learned. For Richard and Michael, their new knowledge led them on to explore related fields in physics and chemistry. Not only do the characteristics of scientific literacy apply to these two authors; they are also technologically literate. For example, they use and understand the designed world, artefacts, systems, and infrastructure. They can identify practical problems, design and test solutions, and evaluate results. Their practical hands-on skills enable them to fix technical problems that are highly complex.

For the two case story authors who were concerned with the environment, a deep understanding of the impact of their chosen field on the local and wider environment was one outcome of their learning. They engaged in discourse with others to further their own understanding of the chosen topic. For Tina, initial learning about bees began to encompass opportunities to discuss bees with other adults, as well as learning from various media. She has widened her knowledge to include the flowers that are attractive to bees and she has joined a social group of like-minded adults. Paulette's specific experience with the geology of her district opened new avenues of learning. She also consulted various media, and found friends who *"did not mind geological conversations"*. This led to her involvement with the University of the Third Age. When controversy arose over the availability of water, the general importance of the water to the local community led to action. Paulette interacted with environmental scientists and expanded her knowledge into several complementary fields.

The group who learned in order to help others also found their own scientific horizons broadening. Paul was willing to consult a museum and pursue the knowledge pathway that ensued. Liz, especially, turned her initial experience into a long quest for further knowledge, largely through the Internet but also by consulting with others. Tiki, already very familiar with some aspects of science, is *"still learning"*. She, too, learns from others through social media in various forms, maintaining a healthy scepticism about information from unconfirmed sources. Kristen finds colleagues an invaluable source of information and they share knowledge through conversations and through social media. Warren needs to *"carry on a process of lifelong learning"* – he consults the media, museums, colleagues, and visitors to his tour sites.

Last, we come to those who have learned science for their work. All of our case stories in this category illustrate continuing learning – Hugh, for example, said *"continued learning and increasing one's knowledge opens so many doors, not only in business, but in personal life"*. Ketan's fascination with scientific information is ongoing and interactive, as is the case for Keith, who continually learns about horticulture and is *"looking at further study"*. Ketan is a sceptic, but remains excited and stimulated not only by those who believe in pseudoscience, but by science in general.

It is our contention that all our authors fulfil the criteria for scientific literacy. In every respect they are lifelong learners, increasing their knowledge over long periods, sometimes decades. It is abundantly clear that our case stories do not simply illustrate isolated learning occurrences. These are stories whose ends are not yet told. The impact of initial exploration to find out something needed and useful has blossomed into a wide appreciation of science more generally. In terms of scientific literacy, however defined, this is a progression that underlines the highly personal yet universally valuable nature of such activity. As time has passed, their degree of science literacy, also, has increased as their knowledge of the science related to their stories has grown. For each person, this knowledge is different. It defies attempts to categorise "desirable" science knowledge as a set of facts that are externally predetermined.

Increasing "Science Capital"

There is another aspect of this learning, however, that refers to the overall science knowledge of an individual and, by extension, a nation. This is the term "science capital" (Archer, Dawson, DeWitt, Seakins, & Wong, 2015; King, Nomikou, Archer, & Regan, 2015). According to Archer et al. (2015):

> The imperative to improve (widen and increase) [science] participation reflects both national economic concerns, namely to ensure a sufficient talent pool and supply of future scientists, and social justice concerns, to promote equity and ensure a scientifically literate general population who can be active citizens within a scientifically advanced contemporary society.
>
> *(p. 922)*

Thus increasing science capital from a national perspective is, in part, an economic imperative. These authors conclude, however, that:

> we feel that the value of science capital lies in its potential to provide a way of understanding the reproduction of inequalities in science participation – and a potential vehicle for dismantling and re-structuring current unequal relations of power: to help create contexts within which other (wider, different) forms of capital might be valued while also re-distributing and sharing out privileged forms of science capital more fairly between social groups. In this respect, we hope that science capital might offer a useful new, or additional, way of promoting social justice within science education.
>
> *(p. 943)*

In sum, an individual's science capital is assessed by a range of measures which include:

- their degree of scientific literacy, as defined above;
- their valuation of science as a component of the life in general;
- an inclination to engage with science-related media;
- a willingness to engage with the ideas and outcomes of science in cultural contexts, e.g., museums;
- having family/friends with scientific knowledge and qualifications;
- knowing people who work in science-related jobs; and
- seeking opportunities to talk to other adults in everyday life about science.

According to the analysis above, every one of our authors increased their science capital in their search for knowledge that they have described. In various ways they consulted the media; they talked with like-minded people. Some visited museums or talked with scientists. All value science in their own lives and the lives of others. Their knowledge has certainly added to their science capital but has not been recorded or measured. How often is this pattern repeated across communities and how, indeed, *might* it be measured? We assert that the very act of learning *in this manner* may be deemed to increase science capital and that some understanding of the extent of informal science learning across communities would enable some estimate of this elusive quality. If the concept of science capital is to gain wide currency, then we need to understand what it means for those who have left the classroom and their years of schooling. In order to do this, however, we must identify those factors that are important for those who wish to further their scientific knowledge and experience, and consider how to encourage their development from the beginning of formal schooling.

Providing Knowledge and Skills to Facilitate Universal Scientific Literacy

It will be impossible to devise a comprehensive curriculum to support the universal attainment of scientific literacy. This would have to accommodate: the diverse approaches that collectively define scientific methodology; the vast variety of phenomena with which science is and will be concerned; the infinite variety of contexts in which that phenomena might be encountered; and the wide divergence of reasons why such knowledge will be sought by individuals.

Schwab (2013 [1970]) wrote a seminal paper about curricular reform, in which he stated that "The field of curriculum is moribund" (p. 591) – a state of affairs that can only be remedied by taking a more practical and less theoretical approach to curricular design. The emphasis on a practical approach and the problems of current curricula that Schwab identified were summarised by Westbury (2013, p. 646) in the context of possible prescriptions for change. Westbury mapped Schwab's ideas within a framework that enables suggestions of what a formal curriculum might achieve through enlightened practical reform.

Westbury separated the problems of framing how science is portrayed in the current curriculum and relevant remedial prescriptions under four headings:

1. *The source of the problem* is the abstraction of scientific concepts that have their origins in assuming that they are "states of mind". In contrast, the remedial prescriptions advocate that science should be framed as "practical problems [that] arise from states of affairs in relation to ourselves".
2. *The outcome of the education* is the assumption that there can be "universal statements which are supposed to be true, warranted, confidence-inspiring", whereas the remedial prescription is "a *decision*, a selection and guide to possible action" (original emphasis).
3. *The curriculum content* currently assumes that there are universal constants and context, "impervious to changing circumstance". The remedial prescription involves "something concrete and particular and treated as indefinitely susceptible to circumstance, and therefore highly liable to unexpected change".
4. *The method* is currently framed within the context of a traditional theoretical inquiry with all its connotations of appropriate data-gathering and interpretations. Again in contrast, the remedial prescription will be one in which "the problem slowly emerges" such that the data search and collection take their direction from this slow emergence.

These steps are not the way that current curricula represent the scientific process and achieving such a design in school curricula will be difficult. Although an extended discussion of curricular change is not our purpose in this chapter, it is striking to note that the solutions suggested in the "remedial prescriptions" closely resemble not only the patterns of scientific research (on which they were based) but the patterns of knowledge seeking evident in our case stories. All involved personal problems that required decisions about action, acceptance of changing circumstances, and a willingness to follow emerging directions as the information unfolded.

Westbury (2013) makes the point that Schwab's work, while problematic in the light of political and educational realities, nevertheless is significant as "an invitation to enquiry/action" (p. 647). Our case story examples of the learning of very diverse aspects of science and technology, undertaken for equally diverse reasons, in many different ways and over a wide range of time frames, underscore the reality of doing science in the real world. They demonstrate very clearly the practical pattern of science inquiry as practiced by non-scientists in search of helpful solutions.

In addition to practical approaches, theories of learning can also contribute to understanding how curricular reform might be achieved. This book has outlined four major theories (classical, constructivist, andragogy, heutagogy) that might, to differing degrees, be relevant to the adult learning of science. Greater knowledge of how the self-directed learning of science and technology can be facilitated, by learners of all ages and circumstances, will be obtained by a closer examination of theories of learning together with accounts of personal educational practice, and the relation between the two. McKeon (1952) has outlined four generic approaches to the relation between theory and practice:

1. The logistic. Theory is then applied to produce the representation of practice;
2. The problematic. Good practice is identified and representational theory is evolved from it;
3. The dialectic. Theory and practice and evolved at the same time and inform each other; and
4. The operational. Only the description of practice is emphasised while the development of theory is neglected.

If we accept Schwab's assertion about the importance of curriculum in educational provision, then we must discount the last of McKeon's relationships. The other three alternatives could be pursued with the hope of insight evolving. Indeed, there is so much work to be done that a start anywhere would be welcome.

Achieving the Goals of Lifelong Learning in Science and Technology

Both Vision II (Roberts & Bybee, 2014) and a Worldly Perspective (Rennie et al., 2012a, 2018b) acknowledge the necessity of locating science learning within the world of the student. They address the need to focus on issues, on critical questioning and problem solving. We assert, however, that science education needs to go even further if it is to provide the kinds of skills and knowledge that our case story authors required. In the words of Feinstein (2011), "We do not need to abandon the … vision of a competent citizenry that can cope with science-related real-life challenges" (p. 170). The outcome of a science education such as we suggest would be a "scientific citizen" – one with all the attributes Rennie and Stocklmayer (2003) set out as desirable goals for how the adult public might be involved in science and technology. These attributes include the following:

- People who feel that science and technology lie within their interest and their personal lives.
- People who feel comfortable about processing relevant scientific information so that their personal areas of interest are well served.
- People who feel that the nation's science is both their property and their responsibility.
- People who are able to access new knowledge in science and technology and understand how it will affect their lives.
- People who feel that their own knowledge and concerns are valued by the scientific community.

(p. 771)

These are the missing attributes that, if present or readily accessible, would have greatly helped our case story authors in reaching their goals. In particular, if people had more practice in the second and fourth outcomes, then their quest for new and relevant knowledge might be easier. These aspects are not directly addressed in much of the literature, except obliquely in phrases like "being able to make informed decisions". What skills do people need to make such decisions and how are they to be informed? These topics must be integral aspects of a science curriculum that caters for lifelong learning and sets up the prospect of building science capital over the ensuing decades.

Changing what happens in schools is a long process. In the meantime, the issue then becomes: how can we assist people who would like to learn more about science and technology for their own immediate purposes? What practical indicators are there that a person who knows some, perhaps only a little, about science and technology is willing to acquire more?

What can be done to increase the participation of adults in learning about science and technology in any way within the activity-type range of "taking a formal course" through to "studying entirely on one's own"? The issue is not one of a lack of agencies that provide support for "learning". For example, an enlightening survey of the "ecology of the science education community" in the United Kingdom identified no less than 17 types of organisation, ranging from universities to broadcasters and hobby clubs, that are involved in making provision to support science education at some level (Falk et al., 2015, p. 153). It was found, however, that these agencies do not actively work together, resulting in a highly disconnected system (p. 164). A greater degree of coherence between these agencies can and should, we assert, be the aim in the future.

We consider in more detail the five goals identified above by Rennie and Stocklmayer (2003).

1 People Who Feel That Science and Technology Lie Within Their Interest and Their Personal Lives

All our authors sought information that was deeply grounded in their personal lives and interests. Initial engagement, therefore, was assured. As we have reported in earlier chapters, however, surveys of the public that seek to measure engagement uniformly report that most people express some interest but are not necessarily "engaged". "Outreach" is one of the central aims of science communication programmes worldwide, whether through the media, museums, or public events, and participation in outreach is increasing. At the same time, it is acknowledged that those who do participate are generally already in some way engaged or willing to be engaged (House of Lords, 2000). The dilemma facing outreach providers is how to reach the large numbers who remain on the fringes of public science. We do not offer a solution here, except to say that a more positive experience of schooling may well assist in this regard. This book is necessarily concerned with those who set out to learn science that is of immediate benefit, and what they might need to help them do so.

This has important cultural overtones, since it is unrealistic to expect strong identification with Western science from people in non-Western countries. Indigenous science, for people in many nations, is close to their cultural and environmental heritage and, as such, is part of that "comfort" expressed in the second goal, below. Identification with gender and culture for all nations is also fundamental. All of the stories in this book, those written by our authors and those from the Moon Diary assignment, describe people who were seeking to locate science and technology within their own interests and their personal lives.

2 People Who Feel Comfortable About Processing Relevant Scientific Information so That Their Personal Areas of Interest Are Well Served

A measure of comfort with accessing information and making use of it to serve an immediate purpose has been identified to depend on several factors. In Chapter 10, we summarised these as the motivation to learn and active engagement in the learning process. This was accompanied by self-efficacy as a learner, including having clear goals and confidence in achieving them, and self-determination, feeling "in control" of the learning, and having a sense of personal responsibility for learning and the ability to choose one's own learning pathway. Further, the learner requires the skills we have previously identified (see Chapter 10): be able to decide what information is needed, access that information and make use of it, and reflect on the value of that information to the task at hand.

These factors are critical for successful processing of information, but they are not enough. The information itself has to be accessible and understandable, as we have stressed in earlier chapters. Too often, relevant information is clouded in impenetrable jargon and, on the Internet, hidden by other "louder" sites which may well contain biased or incorrect information. It will be critical for organisations who seek public access to their information or research to make that information "user-friendly" and to endeavour to ensure that their site is easily found by search engines.

3 People Who Feel That the Nation's Science Is Both Their Property and Their Responsibility

While not directly related to learning about science as an adult, this sense of "ownership" of science is part of self-efficacy in exploring scientific concepts and information. It is, essentially, a characteristic of the confident science "outsider" identified by Feinstein (2011). He stated that:

> People selectively integrate scientific ideas with other sources of meaning, connecting those ideas with their lived experience to draw conclusions and make decisions that are personally and socially meaningful. Engagement with science is a convenient shorthand expression for this process of connecting science with lived experience. It is, critically, the act of an outsider. People do not engage with science by removing themselves from their own social contexts and asking, "what would a scientist do?" They do not, for the most part, seek to become scientific insiders. They remain anchored outside of science, reaching in for bits and pieces that enrich their understanding of their own lives.
>
> *(p. 180)*

In this regard, it is critical that organisations concerned with reaching out to the public, whatever their purpose, include the aim of profiling that nation's science and making it accessible. As we have noted above, this too has cultural overtones, particularly for non-Western countries and including gender issues. How to do this is a matter for current research and debate. Simple measures, however, can be taken immediately: It should be a "given" that science museums represent local as well as international science and portray local "heroes" in an engaging and accessible manner. Textbooks should not focus only on long-dead Western scientists. Unfortunately, these measures are all too rare. In the media, however, the situation is somewhat better, with popular television and Internet personalities gaining public recognition. Stocklmayer and Cerini (2013) have identified a range of people who, in the public eye, are heroes: they range from Nobel scientists to presenters of science on television. All of these heroes are legitimate targets for fostering pride in science achievements. The rise of popular science books is notable, and together with the increasing number of television documentaries and science-related exhibitions, are part of "a remarkable resurgence of interest in science we have witnessed in recent years" (Morris, 2015, p. 1).

Thus this goal requires concerted action on a number of fronts but must be actively acknowledged and, perhaps, adequately funded.

4 People Who Are Able to Access New Knowledge in Science and Technology and Understand How It Will Affect Their Lives

This goal is problematic, since the proponents of new technology and, indeed, new findings in scientific research have not been especially proactive in addressing public concerns ahead of time. Issues relating to genetics, climate science, vaccination, and so on, have been met reactively and too late. It is imperative to draw the public into the science and, if necessary, the "debate", sufficiently early to discuss all aspects of forthcoming technologies both positive and negative in an honest and open manner. We often hear that "this science is too difficult to explain", but this is nonsense. The public does not need to know every detail of the science, but it is critical to convey the expected impact it may have on ordinary lives. It is also important, as has been often stated, not to exaggerate such impacts. Medicine and the environment have both become frequent casualties of "spin".

Finding useful information is, as we have noted, often difficult, particularly on the Internet. We reiterate here that information providers must not only make the information readily accessible, but understandable and personal.

5 People Who Feel That Their Own Knowledge and Concerns Are Valued by the Scientific Community

The image of science as something only for "insiders" is ubiquitous. It is this intimidating factor that deters many from a personal scientific journey in later life. It is often the case that experts offer advice and information without "listening" and this makes contextual learning very difficult. The importance of respect for the learning of "others" has been stressed by many authors, especially following the publication of the House of Lords Report (House of Lords, 2000) that recognised the need for scientists to engage with their publics. In the case of formal adult learning, there must be an opportunity for learners to contribute their own knowledge, thoughts, and opinions and to feel they are respectfully heard.

Of course these goals apply no less to formal schooling. It is clear that addressing the building of science capital is a concern that transcends age and circumstance, and must begin with small children and continue life-long. Most of our case stories authors have initiated voyages of discovery which have been fruitful and rewarding, but it would have been so much more satisfactory if they had been better equipped for their journey.

References

Aikenhead, G. (2006). *Science education for everyday life*. New York: Teachers College Press.

Archer, L., Dawson, E., DeWitt, J., Seakins, A., & Wong, B. (2015). Science capital: A conceptual, methodological and empirical argument for extending Bourdieusian notions of capital beyond the arts. *Journal of Research in Science Teaching*, 52(7), 922–948.

Association of American Colleges & Universities. (n.d.). *What is a liberal education?*Association of American Colleges & Universities. Retrieved from https://www.aacu.org/leap/what-is-a-liberal-education

Donnelly, J. F. (1992). Technology in the school curriculum: A critical bibliography. *Studies in Science Education*, 20, 123–156.

Falk, J. H., Dierking, L. D., OsborneJ., Wenger, M., Dawson, E., & Wong, B. (2015). Analyzing science education in the United Kingdom: Taking a system-wide approach. *Science Education*, 99, 145–173.

Feinstein, N. (2011). Salvaging science literacy. *Science Education*, 95(2), 168–185.

Feinstein, N., Allen, S., & Jenkins, E. (2013). Outside the pipeline: Reimaging science education for non-scientists. *Science*, 340, 314–317.

Felder, R. M., & Brent, R. (2016). *Teaching and learning STEM. A practical guide*. San Francisco, CA: Jossey-Bass.

Fensham, P. J. (1985). Science for all: A reflective essay. *Journal of Curriculum Studies*, 17, 415–435.

Fensham, P. J. (2008). Science education policy-making: Eleven emerging issues. UNESCO. Retrieved from http://unesdoc.unesco.org/images/0015/001567/156700e.pdf

Gardner, P. L. (1994). Representations of the relationship between science and technology in the curriculum. *Studies in Science Education*, 24, 1–28.

Goodlad, J. (1994). Curriculum as a field of study. In T. Husen & T. N. Postlethwaite (Eds.), *The International Encyclopedia of Education* (pp. 21–86). Oxford, UK: Pergamon Press.

Gribble, S. J., & Rennie, L. (2012). Focus on literacy: Linking language and horticulture at Seaview Community School. In L. Rennie, G. Venville, & J. Wallace (Eds.), *Integrating science, technology, engineering, and mathematics: Issues, reflections and ways forward* (pp. 45–53). New York: Routledge.

Habermas, J. (1971). *Knowledge and human interests* (J. Shapiro, Trans.). Boston, MA: Beacon Press.

House of Lords. (2000). *Report of the Select Committee on Science and Society*. London: House of Lords.

Jenkins, E. W. (1994). History of science education. In T. Husen & T. N. Postlethwaite (Eds.), *International encyclopedia of education* (pp. 3–40). New York: MacMillan.

King, H., Nomikou, E., Archer, L., & Regan, E. (2015). Teachers' understanding and operationalisation of "science capital". *International Journal of Science Education*, 37(18), 2987–3014.

McKeon, R. (1952). Philosophy and action. *Ethics*, 62(2), 79–100.

Morris, A. (2015). *Getting to grips with science: A fresh approach for the curious*. London: Imperial College Press.

Paige, K., Caldwell, D., Elliott, K., O'Keeffe, L., Osborne, S., Roetman, P., … Gosnell, S. (2018). *Fresh water literacies: Transdisciplinary learning for place and eco justice*. Adelaide, Australia: University of South Australia.

Rennie, L. J. (2011). Blurring the boundary between the classroom and the community: Challenges for teachers' professional learning. In D. Corrigan, J. Dillon, & R. Gunstone (Eds.), *The professional knowledge base of science teaching* (pp. 13–29). Dordrecht, The Netherlands: Springer.

Rennie, L. J., & Stocklmayer, S. M. (2003). The communication of science and technology: Past, present and future agenda. *International Journal of Science Education*, 25, 759–773.

Rennie, L., Venville, G., & Wallace, J. (2012a). *Knowledge that counts in a global community: Exploring the contribution of integrated curriculum*. New York: Routledge.

Rennie, L., Venville, G., & Wallace, J. (Eds.). (2012b). *Integrating science, technology, engineering, and mathematics: Issues, reflections and ways forward*. New York: Routledge.

Roberts, D. A. (2007). Scientific literacy/science literacy. In S. K. Abell & N. G. Lederman (Eds.), *Handbook of research on science education* (pp. 729–780). Mahwah, NJ: Lawrence Erlbaum.

Roberts, D. A., & Bybee, R. W. (2014). Scientific literacy, science literacy, and science education. In N. G. Lederman & S. K. Abell (Eds.), *Handbook of research on science education* (Vol. 2, pp. 545–558). New York: Routledge.

Schwab, J. J. (2013). The practical: A language for curriculum. *Journal of Curriculum Studies*, 45(5), 591–621. (Original work published in 1970)

Stocklmayer, S. M., & Cerini, B. (2013). The importance of informal learning in science for innovation education. In L.V. Shavinina (Ed.), *The Routledge international handbook of innovation education* (pp. 372–384). London: Routledge.

Stocklmayer, S. M., Rennie, L. J., & Gilbert, J. K. (2010). The roles of the formal and informal sectors in the provision of effective science education. *Studies in Science Education*, 46, 1–44.

van den Akker, J. (1998). The science curriculum: Between ideals and outcomes. In B. J. Fraser & K. G. Tobin (Eds.), *International handbook of science education* (Vol. 1, pp. 421–448). Dordrecht, The Netherlands: Kluwer.

Westbury, I. (2013) Reading Schwab's "The Practical" as an invitation to a curriculum enquiry. *Journal of Curriculum Studies*, 45(5), 640–651.

ABOUT THE AUTHORS

Léonie J. Rennie is Emeritus Professor of Science and Technology Education at Curtin University. Always interested in how and why people learn, her research interests focused on learning in science and technology, particularly outside of school, and integrated STEM curricula, resulting in over 200 refereed publications. She is first co-author of *Knowledge that counts in a global community: Exploring the contribution of integrated curriculum* and co-editor of *Integrating science, technology, engineering, and mathematics: Issues, reflections and ways forward* (both with Grady Venville and John Wallace). She received the Distinguished Contributions to Science Education through Research Award from National Association for Research in Science Teaching in 2009.

Susan M. Stocklmayer AO is Emeritus Professor of Science Communication at The Australian National University (ANU). She is an expert on alternative conceptions and informal learning and has published extensively in the fields of science communication and science education. She was the co-founder of the Australian National Centre for the Public Awareness of Science at the ANU, which has had intensive involvement in outreach for over 20 years. Until 2015, she directed the Centre. Sue has given teacher workshops in creative science teaching for many decades and on all five continents. She was made a Member of the Order of Australia, followed by an Officer of the Order (AO) in 2015, for services to Science Communication and Education.

John K. Gilbert has a background in teaching prior to moving to an academic career. He became Professor of Science Education at The University of Reading in 1988, then Professor Emeritus in 2005. His initial research interests in children's "alternative conceptions" evolved into a focus on "models and modelling" and most recently to both "visualisation and explanation" and "science communication". During 1989 to 2017, he was Editor-in-Chief of the *International Journal of Science Education*, and is currently Co-Editor (with Sue Stocklmayer) of the *International Journal of Science Education, Part B.* In 2001 he received the National Association for Research in Science Teaching award for Distinguished Contributions to Science Education through Research.

INDEX